I would be very honored (and appreciative) if you took the time to leave me a positive rating on Amazon UK.

If you were in anyway dissatisfied with our service or the product that you received, please notify me immediately. I can be reached at elitedigitaluk@sbcglobal.net , I will do whatever it takes to make you happy. Your total satisfaction is my goal.

I strive for 100% customer satisfaction and a 5 Star Feedback Rating. Please don't leave a 4 Star (or lower) rating, as that is calculated as a negative by Amazon and it really hurts my business. If for any reason you do not feel that I deserve a 5 Star Rating, please contact me before leaving any rating.

Thank you again for giving me this opportunity to service your media needs.

Best wishes and kind regards,
Rick Sterling: CEO - EliteDigital UK
elitedigitaluk@sbcglobal.net

"EliteDigital your best source for hard to find items"

Your Order:

AB-12855029b Federal Drug Control: The Evolution of Policy and Practice by Erlen, Jonathon

Rachel Lart
School for Policy Studies
8 Priory Rd
Bristol BS8 1TZ
GREAT BRITAIN

Buyer Email: raelart@bigfoot.com
Order #: 026-0853214-1481120

Dear Rachel Lart,

Thank you very much for your purchase and I do sincerely hope you enjoy your item.

If there is a problem with your order, please email me before returning an item.

Jonathon Erlen, PhD
Joseph F. Spillane, PhD
Editors

Federal Drug Control
The Evolution of Policy and Practice

Pre-publication
REVIEWS,
COMMENTARIES,
EVALUATIONS . . .

"This is a valuable examination of the great turning points in American drug policy at the federal level. Our present policies can be traced and understood through careful historical analysis. The contributors to and editors of this collection offer to present-day policymakers a stimulating account of the American style of confronting the drug problem."

David F. Musto, MD
Professor of the History of Medicine,
Yale University;
Co-author, *The Quest for Drug Control, 1963-1981*

"Is drug abuse a medical or a criminal problem? The answer to that question is rooted in the history of efforts to deal with drug control and illicit drug use. The articles in this book provide insightful analysis and a meaningful context in which to fully address one of the most pressing issues of our time. More than just a history of drug policy in the United States, *Federal Drug Control* will provide crucial answers for a broad range of professionals, from health care providers to social workers and policymakers. The articles have been coordinated masterfully by Drs. Erlen and Spillane into a comprehensive examination of the development of drug-control policy in the United States that goes beyond a dry compilation of legislation and law to discuss the personalities behind the policies that have defined America's approach to drug abuse. Well researched and authoritative, this is an important book."

Michael A. Flannery, MLS
Associate Professor
and Associate Director
for Historical Collections,
University of Alabama
at Birmingham

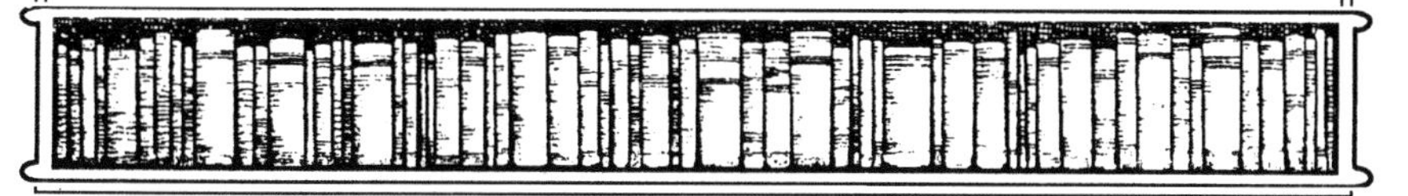

More pre-publication
REVIEWS, COMMENTARIES, EVALUATIONS . . .

"This book's editors and contributors show that the issue of U.S. drug-control policy reached critical mass in the late nineteenth and early twentieth centuries. The chapters, written by scholars in early to mid-career, offer fresh insights into old controversies, such as the Supreme Court's rulings on narcotic maintenance or the motives behind the 1937 Marihuana Tax Act. They also expand on neglected subjects, such as the ethical and legal dilemmas facing pharmacists who refilled prescriptions. *Federal Drug Control* summarizes and updates U.S. drug-control history, while tying it to the larger story of international regulation—a story that seems to have become Sisyphean with the proliferation of new drugs and sources of supply."

David T. Courtwright, PhD
Professor of History,
University of North Florida;
Author, *Dark Paradise: A History of Opiate Addiction in America; Forces of Habit: Drugs and the Making of the Modern World*

Pharmaceutical Products Press®
An Imprint of The Haworth Press, Inc.
New York • London • Oxford

NOTES FOR PROFESSIONAL LIBRARIANS AND LIBRARY USERS

This is an original book title published by Pharmaceutical Products Press®, an imprint of The Haworth Press, Inc. Unless otherwise noted in specific chapters with attribution, materials in this book have not been previously published elsewhere in any format or language.

CONSERVATION AND PRESERVATION NOTES

All books published by The Haworth Press, Inc. and its imprints are printed on certified pH neutral, acid free book grade paper. This paper meets the minimum requirements of American National Standard for Information Sciences-Permanence of Paper for Printed Material, ANSI Z39.48-1984.

Federal Drug Control
The Evolution of Policy and Practice

Patricia — To my dear friend — Love, Rebecca

PHARMACEUTICAL PRODUCTS PRESS®
Pharmaceutical Heritage:
Pharmaceutical Care Through History

Mickey C. Smith, PhD
Dennis B. Worthen, PhD

Senior Editors

Laboratory on the Nile: A History of the Wellcome Tropical Research Laboratories by Patrick F. D'Arcy

America's Botanico-Medical Movements: Vox Populi by Alex Berman and Michael A. Flannery

Medicines for the Union Army: The United States Army Laboratories During the Civil War by George Winston Smith

Pharmaceutical Education in the Queen City: 150 Years of Service, 1850-2000 by Michael A. Flannery and Dennis B. Worthen

American Women Pharmacists: Contributions to the Profession by Metta Lou Henderson

A History of Nonprescription Product Regulation by W. Steven Pray

A Social History of Medicines in the Twentieth Century: To Be Taken Three Times a Day by John K. Crellin

Civil War Pharmacy: A History of Drugs, Drug Supply and Provision, and Therapeutics for the Union and Confederacy by Michael A. Flannery

Federal Drug Control: The Evolution of Policy and Practice edited by Jonathon Erlen and Joseph F. Spillane

Federal Drug Control
The Evolution of Policy and Practice

Jonathon Erlen, PhD
Joseph F. Spillane, PhD
Editors

Pharmaceutical Products Press®
An Imprint of The Haworth Press, Inc.
New York • London • Oxford

Published by

Pharmaceutical Products Press®, an imprint of The Haworth Press, Inc., 10 Alice Street, Binghamton, NY 13904-1580.

Cover design by Lora Wiggins.

Library of Congress Cataloging-in-Publication Data

Erlen, Jonathon, 1946-
Federal drug control : the evolution of policy and practice / Jonathon Erlen, Joseph F. Spillane.
p. cm.
Includes bibliographical references and index.
ISBN 0-7890-1891-8 (Case : alk. paper)—ISBN 0-7890-1892-6 (soft : alk. paper)
1. Drugs of abuse—Law and legislation—United States. I. Spillane, Joseph F. II. Title.
KF3885 .E75 2004
344. 73'04233—dc22

2003016037

CONTENTS

ABOUT THE EDITORS

Jonathon Erlen, PhD, is Assistant Professor of Behavioral and Community Health Sciences at the Graduate School of Public Health at the University of Pittsburgh. Dr. Erlen is in charge of the history of medicine programming for the University of Pittsburgh School of Medicine and teaches four courses in the history of medicine and health care. He received his BA in history from Indiana University, and his MA and PhD in history from the University of Kentucky. Dr. Erlen is co-editor of *The Ladies Dispensatory* and *The Skillful Physician,* and author of *The History of the Health Care Sciences: A Selected Annotated Bibliography, 1700–1980.* He is a contributing editor to fourteen journals and databases.

Joseph F. Spillane, PhD, is Associate Professor in the Center for Studies in Criminology and Law, and in the Department of History at the University of Florida in Gainesville. He received his BA in history from Gettysburg College, and his MA and PhD in history from Carnegie Mellon University. He is author of *Cocaine: From Medical Marvel to Modern Menace in the United States, 1884–1920,* which won the 2001 Addiction book award from the Society for the Study of Addiction. Dr. Spillane has lectured extensively on the history of American drug control.

CONTRIBUTORS

Rebecca Carroll, PhD, is Professor of Communication at Saint Mary's College of California. Dr. Carroll received a BS in English from Clarion University of Pennsylvania, an MA in English from The Pennsylvania State University, and a PhD in rhetoric and communication from the University of Pittsburgh. Her dissertation is titled "The Rhetoric of Harry J. Anslinger, Commissioner of the Federal Bureau of Narcotics, 1930-1962." She has presented many invited lecturers on the history of drug-related topics and in 2000 published "Drug Treatment Policies," an entry in *The Encyclopedia of Criminology and Deviant Behavior,* for Taylor and Francis Publishers.

William B. McAllister, PhD, is Research Historian at the U.S. Department of State in Washington, DC. Dr. McAllister received his BA in history from Oklahoma State University and MA and PhD degrees in history from the University of Virginia. He is the author of *Drug Diplomacy in the Twentieth Century: An International History* (London and New York: Routledge, 2000) and has written and lectured widely on issues involved with the history of international drug relationships.

John P. Swann, PhD, is Historian, U.S. Food and Drug Administration. Dr. Swann received his BA in history and chemistry from the University of Kansas and his MS in the history of pharmacy and PhD in pharmacy and history of science from the University of Wisconsin, Madison. He received the George V. Doerr Memorial Fellowship from the American Foundation for Pharmaceutical Education. His book, *Academic Scientists and the Pharmaceutical Industry* (Baltimore: The Johns Hopkins University Press, 1988), received the Edward Kremers Award from the American Institute of the History of Pharmacy.

Foreword

Significant efforts by the federal government to control the use of illicit drugs in the United States began with the passage of the Harrison Narcotics Act in 1914, although as Joseph Spillane shows in Chapter 1, the roots of federal drug policy can be traced back even further in time. Nearly a century has passed since the passage of the Harrison Narcotics Act, yet the problem of illicit drug use is still very much with us, and substantial controversy continues about what our nation's drug policy should be.

This book provides a broad and thorough overview of the evolution of federal drug policy in the United States, thus helping to place the problem and the policy controversy in an appropriate historical context. For this reason, it will be of interest not only to historians but to social scientists, policymakers, and others concerned with the issue of substance abuse and efforts to regulate the distribution and use of illegal drugs. It should also prove useful in the classroom in a variety of courses.

The book brings together the contributions of a number of knowledgeable scholars, but the individual chapters are woven into a whole, so that the final result is more of a unified account of the evolution of drug policy than a series of disparate essays. The book especially emphasizes the formative period through the 1960s, although attention is also devoted to more recent history, especially in the concluding chapter. *Federal Drug Control* is a welcome contribution to the literature on the history of efforts to control substance abuse.

John Parascandola, PhD
Historian, U.S. Public Health Service

Preface

The federal response to the ever-growing crisis concerning illicit drugs has evolved piecemeal, without a consistent guiding hand. Many individuals have played major roles trying to impose their vision on this complicated situation over the past 125 years. The central issue has been and remains the same: Which of two drastically different approaches should be used by local and federal governments to try to contain illicit drugs? Should the government rely on educational and treatment programs or turn to the criminal justice system? This debate continues into the twenty-first century.

This edited book provides an overview of the evolution of governmental attempts to control illicit drugs in the United States from the last decades of the nineteenth century to 2001. Special attention is focused on key pieces of federal legislation that constructed the federal drug regulatory machinery and the Supreme Court cases that interpreted these laws and their implementation. One chapter presents the story of international attempts at control of illicit drugs from 1880 into the 1990s. Although a number of individuals have been part of America's drug control efforts, Harry J. Anslinger, who ran the Federal Bureau of Narcotics (FBN) for more than three decades, is a prominent figure in this book. The contributors trace the internal tensions between the pro-treatment and pro–criminal justice factions throughout the twentieth century, discussing the difficult choices that have and continue to be made in this ongoing debate.

Joseph Spillane, co-editor of this volume, discusses in Chapter 1 the nineteenth-century origins of the illicit drug controversy, beginning with the rise of large-scale pharmaceutical firms and many smaller drug companies immediately after the Civil War. By 1900 public concern about potential abuse of opiates and cocaine was growing. Although physicians and druggists attempted to limit access to these addictive drugs and some state legislatures passed initial drug control laws, the ease of access to these drugs raised serious public fears. America's acquisition of the Philippines as a result of the 1898 Spanish American War and the concerns expressed at several international drug conferences further heightened the growing awareness that the federal government needed to take action to control America's drug supply. These and other factors during the years 1900 to 1914 led to the passage of the 1914 Harrison Narcotics Act, which required dealers in narcotics to pay a tax on them and to register with the federal government. Considerable confusion followed for several years over the implementation of this law, as physicians and druggists were unsure of their new relationship with the federal government in providing narcotics to the public. This law also had a major impact on all those who used narcotic drugs, both the legitimate, respectable addict and the new breed of "drug fiends."

In Chapter 2 Spillane discusses the events in 1919 that drastically reshaped the implementation of the Harrison Narcotics Act, shifting the approach from public health maintenance to strong police enforcement. The 1919 Supreme Court decisions *U.S. v. Doremus* and *Webb et al. v. United States* greatly strengthened the federal authorities' punitive powers against physicians who were providing drugs to their addict patients, thus making the maintenance approach to drug control difficult if not impossible to continue. Other Supreme Court decisions in the 1920s gave even more power to the criminal justice system. In that decade, several individuals led the fight over which approach to drug control efforts would prevail. Levi Nutt, Lawrence Kolb, Richard Pearson Hobson, and A. G. DuMez led the two competing groups in these debates, with the pro–police power side in firm charge by 1930. The author also discusses the actual law enforcement practices of federal and local officials in the 1920s, including the issue of police/physician relations and the difficulties encountered by police in trying to infiltrate the growing illicit drug trade.

Rebecca Carroll's two chapters, based on her doctoral dissertation, survey the career of Harry J. Anslinger and his long-lasting impact on the American drug scene. In Chapter 3, Carroll presents a biographical overview of Anslinger's life and pre-FBN career. Anslinger became the FBN's first commissioner in 1930 and worked with a zealot's attention to detail in building his federal agency, always striving to monopolize all federal drug control activities and to expand his bureau's jurisdiction. He created an effective working relationship with J. Edgar Hoover, director of the Federal Bureau of Investigation, which proved beneficial to both men. Anslinger developed close ties with influential congressional leaders, thus assuring passage of the 1937

Marihuana Tax Act and the 1951 Boggs Act, both of which extended his agency's powers. He made very effective use of the national media, using horror stories of drug abusers attacking innocent victims, to further his personal antidrug agenda that relied on harsher laws providing longer prison sentences and de-emphasizing or totally ignoring all education and treatment alternatives for illicit drug use.

In Chapter 4, Carroll chronicles Anslinger's victorious efforts in the 1950s to eliminate all other voices from the drug control discussion. He continued his mastery over Congress, getting passage of the Narcotic Control Act of 1956, which featured much harsher penalties for drug-related crimes, including the potential of the death penalty for anyone convicted of selling heroin to someone under the age of eighteen. Anslinger had virtual control over federal drug control policies and was powerful enough to totally dismiss the pro-treatment views of the American Bar Association-American Medical Association Joint Committee. He continued to use the media, including *The New York Times* and the radio program *Monitor,* to scare the American public into accepting his punitive approach to the handling of illicit drugs. Anslinger's legacy is clearly visible in today's federal prison population, the continuing belief that marijuana is the gateway drug which leads to more severe addiction, and the opposition of funding for drug control educational and treatment programs.

In Chapter 5, John Swann discusses the increasing battles between the federal government and the pharmacy profession over pharmacists' rights to refill prescriptions of potentially addictive drugs without a physician's authorization in the middle third of the twentieth century. Both the 1914 Harrison Narcotics Act and the 1938 Food, Drug, and Cosmetic Act had a significant impact on the professional pharmacist's activities. By the 1930s, the U.S. Food and Drug Administration had become very concerned about pharmacists filling prescriptions for dangerous drugs to treat venereal diseases without a physician's orders. This concern, along with labeling issues and confusion over what was a prescription drug as opposed to an over-the-counter drug, led to the passage of the Durham-Humphrey Amendment in 1951, clarifying the definition of prescription drugs and requiring physician authorization for refills. This amendment effected basic changes in patterns of self-medication by the American public and the daily practice of the pharmacy profession. In the same period, the FDA dedicated more time to the interdiction of illegally sold amphetamines and barbiturates than all other pharmaceutical violations of the 1938 act combined.

Many historical studies on American drug issues neglect the significant impact of international efforts to control illicit drugs. In Chapter 6, William McAllister covers this topic from 1880 into the 1990s. Beginning with concerns over the "Great Game" in East Asia in the late nineteenth century, American politicians have realized the close connection between American drug control interests and their international context. Drug-related legislation, both international and national in scope, has been based primarily on economic, political, religious, and social factors rather than medical concerns. American drug control efforts have paralleled those of other Western nations during the twentieth century. This shared approach to drug-related concerns led to the participation of the United States in a number of international drug control agreements, including the 1912 Hague Opium Convention, the 1961 Single Convention, the 1971 Psychotropic Convention, and the 1988 Illicit Traffic Convention. Harry J. Anslinger played a key role as U.S. representative to numerous international drug conferences from the 1930s into the 1960s. His efforts to ensure a supply-control approach to international drug issues were not accepted by the international community. Today American drug policy experts must plan their strategies around the realities of the global drug market.

In the final chapter, Spillane accepts the challenging task of surveying the increasingly complex world of American drug control efforts in the post-Anslinger era. The author provides a bibliographic guide to many of the major primary and secondary sources from America's drug control literature during the past four decades. These forty years have seen an explosion in the use of illicit drugs worldwide. Social science literature, beginning in the 1960s, has led the criticism of Anslinger's attempts at drug control, reopening the debate between pro-education/treatment and pro–criminal justice approaches to this topic. President Nixon (1969-1974) drastically altered Anslinger's tactics on illicit drugs, and the Comprehensive Drug Abuse and Control Act of 1970 restructured America's federal drug control machinery. Under President Reagan's leadership, throughout most of the 1980s the United States embarked on a new war on drugs, increasing federal spending and creating more administrative machinery. Numerous studies during the past four decades have examined the emergence of new illicit drugs, such as the psychoactive drugs, provided in-depth history of specific drugs, most notably marijuana and cocaine, reviewed the international aspects of the drug trade, and described the broader issues involved in the evolution of drug use and drug policies. This chapter provides a guide for those seeking information on these complex, ever-changing issues.

A list of URLs is provided following the text in each chapter. These listings provide access for scholars and students to the full text of relevant federal laws, Supreme Court decisions, and congressional hearings discussed in each chapter.

The editors and authors of this book wish to thank Dennis Worthen, Lloyd Scholar, Lloyd Library, for all of his invaluable support of our efforts, and The Haworth Press for its administrative assistance with this project. This book is dedicated to our longtime colleague, mentor, and friend, James Harvey Young, senior historian of American health care quackery.

Jonathon Erlen

Chapter 1

The Road to the Harrison Narcotics Act: Drugs and Their Control, 1875-1918

Joseph F. Spillane

After several years of debate and negotiation, the U.S. Congress passed the Harrison Narcotics Act, which went into effect on March 1, 1915.[1] Designed primarily to control the nonmedical sale and use of opiates and cocaine, the act limited distribution to those doctors, druggists, manufacturers, and wholesalers who registered with the federal government, paid a tax, and kept detailed records. One of the cornerstones of federal drug legislation until 1970, the Harrison Narcotics Act effectively ushered in an era of national drug control.[2] The story of drug control in the United States began decades earlier, however, with the emergence of concerns about the nature and effect of certain substances. With the growing capacity of the drug industry to deliver opiates and cocaine to the American consumer in attractive and inexpensive forms, communities around the country began a process of defining what constituted "legitimate" drug use and sale. By the turn of the century, the process of state and local governments translating those standards into legal rules was well underway. By 1914 even the federal government had considerable experience with drug regulation. It would be reasonable to say, then, that the passage of the Harrison Narcotics Act represents as much a culmination as a new beginning. This chapter explores three critical early developments: the transformation of the drug industry, the rise of antidrug sentiment, and the early efforts at legal controls at local, state, and federal levels.

DRUGS AND DRUG DISTRIBUTION IN LATE NINETEENTH-CENTURY AMERICA

For much of the nineteenth century the American pharmaceutical trade had remained much as it had long been, characterized by many small operations that supplied botanical preparations to physicians. Drug firms operated reactively, supplying those remedies favored traditionally in medical practice. The ability to stock as many plant extracts, tinctures, and powders as possible stood as the standard of excellence for drug suppliers.

After the Civil War, drug makers began to show some of the earliest signs of the changes that would transform the industry over the next century and make the drug enterprise one of the nation's most important areas of manufacturing. Along the way, the industry would sharpen the distinction between the two major groups of drug manufacturers. The first group included the so-called ethical drug firms, who identified themselves as such on the grounds that they supplied drug products exclusively for medical purposes. Indeed, the ethical firms took great pains to publicize the fact that they did not make direct advertising appeals to the general public, but confined their sales pitches to persuading doctors and druggists of the superiority and reliability of their brands. By the turn of the century this ethical business had come to be dominated by a relatively small number of firms. No more than a dozen ethical pharmaceutical firms accounted for most of the drug products supplied to the nation's doctors, and most of these companies were still dominated by the men who founded them: Edward Robinson Squibb, Eli Lilly, John Uri Lloyd, A. R. L. Dohme, H. K. Mulford, Wallace Abbott, and others. Nevertheless, the corporate structure of these firms had evolved considerably from their founding. Expanding from regional interests to national markets, the drug industry was aided immeasurably by the development of new manufacturing technology. New pill-making machines sped up the manufacturing process and made the production of large quantities easier. The resulting product proved more reliable and efficacious, with greater dosage standardization, qualities appreciated by physicians. Research conducted by Parke, Davis led to the standardization of fluid drug extracts, long a staple of the drug business, through new manufacturing processes.[3]

Changes in corporate structure and technology alone would have remade the drug business, but the full character of the modern industry emerged with a new and aggressive approach to drug development, research, marketing, and promotion. If any one figure is entitled to the credit for having brought about these changes, it may well be George S. Davis, partner of Hervey Parke, and general manager of Parke, Davis and Company during the critical decades of the 1880s and 1890s. It was Davis who recognized most clearly that the capacity of the industry might be turned not only to the production of those products that physicians demanded, but also to the development of their own drug products, sparking medical interest through the publication of favorable research results.[4]

Cocaine, introduced into widespread medical use in the fall of 1884, was among the first drugs to benefit from this newly aggressive approach to drug development and distribution. Although medical experimentation had initiated the burst of interest in cocaine, drug makers such as Parke, Davis quickly reprinted favorable studies in their own medical and pharmaceutical journals. The company also pursued its own investigations into potential therapeutic uses of cocaine. When the company decided to launch a line of coca cigars, for example, it employed Philadelphia physician Francis E. Stewart to conduct experiments on their stimulant and tonic effects. Stewart duly published an article, "Coca-Leaf Cigars and Cigarettes," in the *Philadelphia Medical Times*.[5]

If George Davis pushed the ethical drug business closer to its modern attitude, he also caused a reaction that laid the groundwork for government regulation of the pharmaceutical industry. The health professions, while appreciative consumers of both the drugs and drug information generated by industry, worried that the physician and pharmacist would ultimately become mere slaves to the pharmaceutical trade. Above all, physicians began to suspect that the corporate desire for profit would overwhelm care for public health and safety. How were physicians to keep up to date on the newest drug products without simply relying on the claims of the manufacturer? In the hands of the naive consumer or the careless doctor, the latest products of the drug laboratory might be abused, as the president of the American Association for the Study of Inebriety observed in 1902:

> New drugs are constantly placed on the market and we can recall a time when bromism, chloralism, cocainism were uncoined

> words, because the bromides, chloral hydrate, cocaine had not yet emerged from the laboratory of the manufacturing chemist and their danger to heart and brain was yet to be experienced in actual practice. And let it here be noted—in parenthesis—that too often the child of the laboratory is a product not only for good but frequently for evil, so that we might almost say, "Would that it had never been born."[6]

Alongside the ethical drug manufacturers were an enormous number of firms selling so-called patent medicines. Patent medicines were not actually patented; rather, they were protected by trademark and were consequently referred to by some as "proprietary" remedies. The *Pharmaceutical Era*'s "Druggists Directory" listed 5,398 drug manufacturers at the beginning of the twentieth century, nearly all of whom produced patent medicines. Unlike the ethical business, the vast number of patent medicine firms included everything from national companies with extensive product lines to small regional and even local firms dependent on a single product.[7] The patent medicine industry distinguished itself from ethical firms not so much by their products, which sometimes overlapped with those of the ethical industry, but by their choice of targets. Patent medicines were directly, and without apology, aimed at the general public. Much of their advertising decried the overabundance of physicians, when consumers could do just as well through careful self-medication. Mocking the inability of physicians to cure, patent medicines promised sure relief.

Patent medicine makers strove to maintain at least a veneer of respectability, always defending the medical value of their products. The patent medicine industry never endorsed the view that their products might be misused or used for nonmedical purposes, or that they marketed their products with that particular market in mind. Nevertheless, it is true that alcohol, morphine, and later cocaine were among the most important elements in most of the industry's mainstay products. Wholesale drug catalogs of the late 1880s, for example, contain dozens of products containing cocaine, including Coca Malta, Cocalac, Coca Cordial, Peruvian Tonic Bitters, many coca wines, coca soft drinks, and cocaine snuffs and sprays.[8]

Neither patent medicine makers nor ethical manufacturers could do business without courting the support of the final element in the drug distribution system, the retail druggist. In an age when prescription requirements were far from universal, retail druggists possessed a remarkable degree of discretionary control over the nation's drug supply. Druggists had the power to recommend certain products, to make informal diagnoses, and to refuse sales, and some still compounded their own remedies. As *American Druggist* observed, "there is no man engaged in trade who is on quite such intimate terms with his patrons as the retail druggist," making them a primary target for marketing appeals from drug manufacturers of all types.[9]

By 1900, the rise in public concern over the misuse of opiates and cocaine highlighted the enmity between the major interests associated with the drug business. To the ethical drug industry, patent medicines brought both formidable competition and unwanted disrepute to the pharmaceutical industry. To physicians, patent medicines (the "forces of evil," according to the AMA) also represented competition, as well as a direct challenge to their claims of authority over the nation's drug supply. Physicians regarded most retail druggists as willing accomplices to the patent medicine trade, sacrificing professional integrity for huge profits. Among retail druggists, the reaction to patent medicines was more ambiguous. To be sure, no druggist wanted to become merely a vendor of potentially dangerous patent medicines, but neither could most retailers afford to ignore the economic importance of such products to their enterprise. Druggists saved much of their harshest criticism for doctors, frequently pointing out that with respect to the drug habit, "the doctor has created the desire and the druggist has only ministered to the supposed need."[10] As the debate over the control of opiates and cocaine advanced, the central question for each group was not whether the state would ultimately restrict some aspects of distribution and sale, but whose authority would be privileged.

DEFINING THE BOUNDARIES OF LEGITIMACY

Before the development of formal, legal controls over drug distribution, an informal process of defining inappropriate sale and use was well under way. The descriptions most frequently employed to sort out the nature of certain kinds of drug use were "legitimate" and "illegitimate," terms that were hardly precise but had some widely shared meanings. What aspects of a drug's sale or use were illegitimate? The important distinctions typically followed from one of three general areas of concern: medical, professional, and social.

Medical standards for judging drug use sought to define the effects of specific substances on their users. Since morphine and cocaine had a significant role in medical practice, these standards tended to balance both the potential negative effects (toxicity, addiction) and the potential positive effects (therapeutic benefit, pain relief) in making judgments about the legitimacy of certain types of use. Where cocaine, for example, aided surgeons in anesthetizing mucous membranes, the benefit strongly outweighed the minor risk of untoward patient reactions. Where cocaine had only limited therapeutic utility, as in the case of treating nasal catarrh, the benefit weighed against the prospect of patient self-administration of cocaine sprays, leading to habitual use. Finally, where cocaine provided little or no therapeutic benefit, as in the treatment of opiate addiction, the numerous reported instances of opiate addicts becoming dependent on cocaine strongly weighed against this use. Such areas of therapeutic practice declined quickly and notably.[11] To some extent, then, drugs such as morphine and cocaine were judged by a rational-objective standard consistent with that valued by present-day advocates of a medicalization of drug control.[12]

Phrases such as "legitimate medical use" recur throughout twentieth-century drug policy; the Comprehensive Drug Abuse and Control Act of 1970 enshrined the concept in federal legislation by creating drug schedules based on substances' potential for abuse and "accepted medical use" or "recognized medical value." The physician has a considerable legitimating effect on drug use; even many staunch opponents of marijuana decriminalization, for example, appear to favor its availability for legitimate medical purposes. To some extent, the concept of legitimate medical use simply covered any application physicians chose to make, while drug use without the involvement of a physician was de facto illegitimate. In numerous discussions of the drug problem at the start of the twentieth century, commentators attempted to estimate the quantity of drugs going to legitimate use—by which they meant the quantity that was employed by physicians, on the recommendation of a physician, or through self-medication within accepted medical boundaries.

This highlights another important dimension of the process of defining illegitimate use—the professional standards that defined any nonmedical distribution, promotion, or use as illegitimate, dangerous, and inappropriate. This is what one doctor in 1915 meant when he asked, "would it not be better to fight the bad use, and not hinder the good use?"[13] At this level, concerns over the popular use of opium, morphine, and cocaine merged with the broader professional conflict over control of the nation's drug supply. Some drugs, the argument went, were simply too dangerous to be placed in the hands of the general public or, as one physician summarized the view, "The best things in the world are often the hardest to handle. This is true in medicine. The remedies that are most relied upon and do the most good are often the ones that are worst used."[14] The secretary of the Medical Society of the County of New York described the role physicians cast for themselves as that of "protection of the simple-minded and easily duped from the wiles of the drug charlatan."[15]

This dual impulse of concern, for limiting the harms associated with drug use and the desire to regulate access to the drug supply, joined with a third critical impulse: fear of the users themselves. While iatrogenic addiction could be regarded sympathetically as accidental or unfortunate, pleasure use seemed little more than gross self-indulgence. A physician observed considerable sympathy for "the innocent and ignorant formation" of drug habits, but "the vicious user of a drug whose sole excuse is the seeking for new sensations, is a person who does not need protection, but rather restraint by law in order that he may not become a menace to the public weal and a care for the public charities."[16] In addition, the association of specific drugs with particular groups and classes blended antidrug attitudes with prejudice, nativism, and racism. Group identity and pleasure use were often linked together in analyses of the drug problem. W. C. Fowler, health officer for the District of Columbia and a practicing physician for thirty-five years, argued in 1923 that "the term 'fiend' can be very properly applied to a great many of the underworld persons who were originally mentally and morally degenerates . . . many of the addicts began their addiction through a desire to gratify certain sensual pleasures."[17]

The first significant form of illegitimate drug use in America was opium smoking, which appeared to fail all three tests of legitimacy. Chinese immigrants arriving in the western United States popularized opium smoking in the years following the Civil War. Critics concluded that opium smoking had serious, degenerative effects on persons

ate drug use unquestionably influenced use and distribution, although complete control of consumption by doctors and druggists was more hopeful than realistic. In the absence of formal controls, consumers still found numerous outlets for opiates and cocaine. Nevertheless, voluntary restrictions on certain kinds of drug sales began the process of forcing those outlets underground. Doctors, like druggists, recognized that becoming voluntary participants in drug control afforded them an opportunity to reinforce their claims to professional status, and would allow them to influence the content and direction of any future legislation. In 1903, James H. Beal of the American Pharmaceutical Association framed the central issues in much the same terms as the AMA:

> if the druggists of the United States do not resolutely take hold of the regulation of the sale of narcotic drugs . . . they will merely be turning it over to the care of people who are less competent to deal with it than themselves.[26]

EARLY DRUG LEGISLATION

Although the Harrison Narcotics Act was certainly an important milestone in the history of drug control, it hardly represented the first use of the criminal law in this area, even at the federal level. Historians who debate the significance of the Harrison Act still agree that the 1914 legislation merely added a new layer of controls to an already extensive system of state and local laws. Some critics of U.S. drug control policy, however, have preferred to draw a more stark contrast between the pre– and post–Harrison Narcotics Act eras. Rufus King, writing in 1953, was among the most influential:

> Our grievous error was in allowing the narcotics addict to be pushed out of society and relegated to the criminal community. He isn't a criminal. He never has been. And nobody looked on him as such until the furious blitzkrieg launched around 1918 in connection with the enforcement of the Harrison Act.[27]

For all the merits of King's critique of drug prohibition, he failed to appreciate how deep its roots were in American society.

The idea of controlling drug supplies went back at least as far as Britain's Pharmacy Act (1868) that had regulated the sale of opium. In the United States, the first wave of state pharmacy laws in the 1860s and 1870s often set standards for the handling of "poisons" and dangerous drugs in the course of professional practice. Specifically antidrug legislation dates back at least to San Francisco's opium ordinance of 1875. Nevada passed the first state opium law in 1877 and was joined by Arizona, Montana, North Dakota, and South Dakota before 1900. Responding to newspaper accounts of cocaine use in the Chattanooga underworld, the Tennessee legislature considered a comprehensive drug law in 1897 and 1899. In 1901, the law finally passed, banning all sales of cocaine except those on a physicians' prescription.[28] By that time similar legislation had already appeared in Arizona, Arkansas, Colorado, Illinois, Louisiana, Maine, and Montana.[29] By 1914, virtually every state had cocaine laws, and the majority had legislation dealing with opiates as well.[30]

The use of state authority to control the licensing of pharmacists allowed a number of states to wage impressive campaigns against druggists who sold cocaine. In Pennsylvania, the Board of Pharmacy and its crusading vice president, Dr. Christopher Koch Jr., investigated charges of excessive cocaine selling in the state and conducted a drug sweep in Philadelphia that netted more than sixty people. Like many anticocaine laws, regulations embedded within state pharmacy laws targeted illegitimate sales, a concept that could never be defined with precision. The Kentucky pharmacy code prohibited the sale of cocaine for "illegitimate purposes." It was vigorously enforced, despite the failure to define what illegitimate purposes were. In Cleveland, the local druggists' association suspended cocaine sellers from its membership using a clause in its rules that permitted the suspension of members for "gross immorality."[31] It was California's Board of Pharmacy that pushed for passage of that state's aggressive antidrug legislation of 1909 and subsequently enforced its provisions.

Most early legal controls on the state and local levels were designed to regulate, rather than prohibit, certain kinds of drug sales. The earliest, most common laws dealt with opium smoking and typically focused on controlling the spread of opium dens, or confining them to certain areas of the community. More than twenty states and territories had opium den laws before 1900. Ronald Hamowy observed that most had language similar to that of California:

> Every person who opens or maintains, to be resorted to by other persons, any place where opium, or any of its preparations, is sold or given away to be smoked at such place; and any person who, at such place, sells or gives away any opium, or its said preparations to be there smoked or otherwise used; and every person who visits or resorts to any such place for the purpose of smoking opium or its said preparations, is guilty of a misdemeanor.[32]

Cocaine presented a slightly different problem than smoking opium, since physicians continued to employ cocaine in medical practice to some extent. Therefore, state and local regulation attempted to distinguish between legitimate and illegitimate forms of sale and distribution. In New Orleans, a city ordinance did not prohibit the sale of cocaine or medicines containing cocaine in "recognized therapeutic doses," but did prohibit the "combination of the drugs which may now or hereafter be made wherein the ingredient cocaine gives such proportions as to make its deliriant, or intoxicating, effect the main reason for their use." In 1913 New York banned the sale of flake or crystal cocaine, employed in solution by physicians—used by recreational "sniffers"—while leaving unrestricted the sale of large crystal cocaine employed in solution by physicians.[33]

Of course, the mere existence of formal legislation itself says little about the actual conduct of this emerging "war" on drugs. The general view has been that these laws were only sporadically enforced and full of loopholes, and indeed they were. David Musto stated, "no consistent police efforts to enforce these laws appear to have been undertaken."[34] This, too, seems accurate. Despite a lack of empirical evidence, it can safely be said that drug laws were used in highly selective ways. At moments when popular or political pressure demanded "action," police departments or state pharmacy boards might act aggressively, while at other times allowing enforcement to lapse entirely.

Historian Roger McGrath has shown that, in small western mining towns such as Bodie, California, Chinatowns had well-earned reputations as vice districts, and their opium dens were often scenes of disorder and violence. Bodie's saloonkeepers banned addicts from their establishments, and the sheriff in 1881 began enforcing a state law that called for a $500 fine for opium smokers and a $1,000 fine for dealers. This pattern repeated itself across the West and Southwest;

the number of police actions recorded in newspapers such as the *Butte* (Montana) *Miner* and the *Montana Standard* were overwhelmingly taken against Chinese opium dealers. In Tucson, Arizona, the very first local vice law passed (1880) dealt with opium smoking and the operating of opium dens.[35]

Even farther east, opium was often the first drug to be the subject of legal controls and vigorous prosecution. Chicago police arrested hundreds of persons a year for violations of opium smoking laws, with a peak of 534 such arrests in 1901.[36] Chicago police actively regulated the drug trade before 1914. Periodic political pressure to "clean up" certain areas of the city produced well-publicized police sweeps. Reporters, tipped off in advance, gathered to watch the action. In the confusion of Christmas raids on the West Side in 1911, reporters covering the action watched young Walter McKendrick dive head first from the third-story window of Hector Mariner's cocaine flat. In between the periodic roundups, police activity targeted the small-time sellers who operated in the most visible ways and could not afford to pay substantial protection money. Sellers on the street were constantly on the watch for being "pinched" by the police, sometimes developing a habit of selling and walking at the same time to avoid attracting too many addicts at once and drawing unwanted attention to themselves. Because the earliest laws dealt with sales, not possession, sellers caught with drugs would often be charged with disorderly conduct or some other offense. In this way, an arresting officer could take a seller into custody without having to prove that a sale had taken place. As a consequence, only twenty-five criminal cases for cocaine selling were disposed of in the Municipal Court of Chicago between 1908 and 1913, despite the reported arrest of a far greater number of sellers.[37]

Arrest and prosecution of addicts prompted various sorts of state and local institutional arrangements for drug users. The most common was the county jail, where many users could expect to spend short amounts of time. Drugs were bought and sold in all big-city jails; the opiate addict who could not obtain a source of supply went through withdrawal with no medical assistance. In some instances, jails operated quasi-medical "drunk tanks" for the alcoholic and drug addicted. More unusual still, but not unimportant, were various state efforts at confining addicts in public inebriate asylums. Historians Jim Baumohl and Sarah Tracy have shown how California and Massachusetts created public inebriate asylums that managed large numbers of drug addicts.[38] As with the criminal justice system, asylums' interest in dealing with drug users was unpredictable and tentative, but their mere existence reinforces the conclusion that there was no free market in opiates and cocaine in this era. Regulatory efforts may have been sporadic and variable, but they belie the notion of a disinterested state.

THE ROLE OF THE FEDERAL GOVERNMENT

The limitations on federal police power as the nineteenth century came to a close were strikingly apparent to any observer of national government. Although a few commentators on the drug problem were moved to predict that the time was near when the federal government would exert substantive regulatory control over the drug supply, still more questioned whether there would ever be an attempt to impose national control. Indeed, in a legal climate generally hostile to most forms of federal regulation, it seemed an open question whether the imposition of a standard, national authority was even constitutionally permissible. Over the course of the first fifteen years of the twentieth century, events would propel drug control in the direction of the federal government.

First, a coalition of medical and reform interests won a significant victory in helping to secure passage of the Pure Food and Drug Act of 1906. This legislative action, together with the AMA's Council on Pharmacy and Chemistry, led to the first general rules for drug promotion, packaging, and distribution. The Food and Drug Act imposed a national labeling requirement, requiring manufacturers to list the presence of certain drugs in their products. Section 8 of the Act enumerated those drugs: morphine, opium, cocaine, heroin, alpha or beta eucaine, chloroform, cannabis indica, chloral hydrate, acetanilid, or any "derivative or preparation of any such substances." Although the act, administered by Harvey Wiley's Bureau of Chemistry, held out the possibility of dealing with false or misleading claims, initial enforcement focused largely on mislabeling. The Council on Pharmacy and Chemistry, established in 1905, proved a valuable adjunct to Wiley and the bureau. The council's purpose was to provide an official stamp of approval for all drug products; a company risked disapproval if it made any specific therapeutic claims, failed to disclose ingredients, or advertised directly to the public. Like the Bureau of Chemistry, the council gave itself additional weight through an active publicity campaign against drugs that failed their tests. In this campaign, Wiley and the AMA were strongly aided by the work of certain muckraking journalists, the most notable of whom was Samuel Hopkins Adams. His sensational series of articles for *Collier's* magazine began appearing in 1905. In the articles, later reprinted by the AMA, Adams excoriated the patent medicine industry for "doping" the American public and perpetrating what he called "the great American fraud."[39]

The impact of the Food and Drug Act is easy to misinterpret. As James Harvey Young has observed, the meaning of "pure" had evolved somewhat by the time the act was in force. Where pure at one time was nearly always interpreted as meaning labeled with scrupulous accuracy, after 1906 it often meant safe and free from dangerous qualities. In practice, this meant that the addition of cocaine or opiates to a product might render it "impure" despite the presence of strictly accurate labeling.[40] In addition, Section 7 of the act allowed claims of adulteration of food products if the food "contained ingredients which may render such product injurious to health"; cocaine was one such ingredient.

The government's summary in a Section 7 case against the soft drink Celery Cola made it clear that the intent of the prosecution was to demonstrate that cocaine "was a drug which should never be administered in . . . a product which was available for all persons at all times."[41] Clearly, this area of regulation reflected the varied impulse of drug control—grounded in a concern for limiting the harms associated with drug use, but also in a desire to control the drug industry more generally. By the standards of the latter interest, no exceptions were possible—any commercial interest in opiates or cocaine would inevitably threaten the public interest, if not the public health.

Second, a series of international developments placed the federal government in a position of leadership in the drug control arena.[42] When it assumed control over the Philippines following war with Spain in 1898, the United States inherited a substantial and apparently growing opium problem. The Episcopal bishop of the Philippines, together with civil governor William Howard Taft, worked to craft a policy to deal with opium smoking. Their plan, announced in 1904, would have gradually eliminated the legal market for smoking opium over three years; Congress ordered immediate prohibition the following year. The American government quickly expanded its engagement with the opium question to include the issue of opium in China, repeatedly urging the organization of an international conference to deal with the question.

Dr. Hamilton Wright led these efforts. Wright had made his reputation as a disease specialist, but his great passion proved to be drug diplomacy. Working with the U.S. State Department, Wright helped organize a series of international meetings, including the Shanghai Opium Commission in 1909 and the International Conference on Opium at The Hague in 1911-1912. These meetings yielded limited results but helped to capture the attention of American lawmakers.

The requirements of global leadership on the drug question raised for Hamilton Wright the potentially embarrassing lack of domestic legislation. The first attempt to remedy this problem came in the form of legislation banning the importation of opium for smoking. Congress passed the measure early in 1909, in time to present it to the Shanghai Opium Commission as evidence of American commitment to dealing with drug problems at home as well as abroad. The approach of conferences at The Hague in 1911 and 1912 led to further pressure for still more substantive national legislation. From these efforts came the first abortive attempts to pass versions of what would become the Harrison Act.

Wright himself drafted the first model domestic legislation. When Representative David Foster of Vermont introduced the first bill in 1910, it was based on Wright's model of controlling narcotic drugs through taxation. When the House Committee on Ways and Means held hearings, Wright testified at length. The drug industry fiercely opposed the Foster Bill, arguing that its record-keeping requirements were too onerous and the exemptions were not broad enough. Hearings continued into January 1911, featuring some of the most open discussion of the merits of drug control yet heard, but the following month the bill died.

In 1913 Representative Francis Burton Harrison introduced a new federal drug control bill, one that initially seemed much like the Foster Bill. Rep. Harrison was a New York Democrat who had first arrived in Congress in 1903, serving on both the Ways and Means and the Foreign Affairs committees. The Harrison Bill was the final chapter in Harrison's career in Congress; later that year President Woodrow Wilson appointed him governor-general of the Philippines. Indeed, it may well be that Harrison's interests in the Philippines, with its significant opium problem, prompted his initial interest in the drug question at home.[43]

The bill that Harrison introduced was the subject of lengthy negotiations with drug and medical interests. Indeed, various drug interests formed the National Drug Trade Conference in 1913 primarily to deal with the pending legislation. Congress passed the Harrison Narcotics Act on December 14, 1914. Three days later, President Woodrow Wilson signed the bill into law. Section 1 of the act required all dealers in narcotics to register with the federal government and to pay a special annual tax. Moreover, all transactions were to be recorded on official order blanks and kept for two years. Section 2 of the act exempted physicians, in the course of professional practice, from having to comply with the record requirements. Section 6, to mollify the patent medicine interests, exempted products with small quantities of heroin, morphine, and cocaine. Section 8 made possession of the regulated substances presumptive evidence of a violation of the act and placed the burden on the individual to prove that purchases had been legitimately made. Section 9 of the act set the penalties for violations at no more than $2,000 or five years in prison.

The Harrison Narcotics Act has long been regarded as the starting point for American drug control, but in some respects the act was merely a culmination of trends that had begun much earlier. Defining the boundaries of legitimate drug use began with countless individual decisions by doctors, druggists, and others regarding the sale and distribution of opiates and cocaine. Although the informal standards of the public and the health professions were never universal or precise, the first generation of state and local controls represented an attempt to formalize these general attitudes. Highly variable in both content and enforcement, the prevalence of these experiments in drug control suggests a highly regulated "free market" in opiates and cocaine. These early controls also reveal multiple, and sometimes contradictory, impulses: earnest concern over the impact of drugs on health and well-being; eagerness to control the nation's drug supply; desire for global leadership; and, perhaps above all, fear of the degenerate drug "fiend" and the "vicious" addict.

GREAT EXPECTATIONS AND UNANSWERED QUESTIONS

If the Harrison Narcotics Act was a culmination of years of drug control efforts, it was also a new beginning. For the first time the national government appeared poised to use the substantial power of federal law to bring order out of the multiplicity of state legislation dealing with the trade in opiates and cocaine. The 1915 *Literary Digest* spoke of a new "AntiDrug War," and noted "foes of the drug evil are welcoming the arrival of a new howitzer battery, so to speak."[44] The *New York Herald* looked forward to "the possibility of great reform in the serious abuse of dangerous drugs." The *Outlook* observed that the Harrison Narcotics Act "marks a notable change in the direction of greater National control of what has come to be a crying National evil."[45]

Yet, for all the optimism, most interested observers remained uncertain just how the new law would work in practice. Many of the most critical questions remained unanswered. Was the act merely a taxation measure designed to confine the business in narcotics to legitimate drug sellers and physicians, or would there be some regulation of legitimate practice? If the intent was to regulate professional activity, how strict would the oversight be, and was this constitutional? Would opiate addicts be allowed to continue to receive their regular supply through physicians? Nearly everyone contributed to the negotiations over the act. Until passage, doctors, nurses, pharmacists, patent medicine makers, and the major pharmaceutical companies battled each other and the government over the provisions of the law. After passage, few knew for certain where they stood.

The early months after enactment featured much confusion, even on the part of the federal government. In the first four months the Harrison Narcotics Act was in force, a startling 5,085 violations were reported. Over 4,000 of these violations were deemed technical in nature, and reflective of "misunderstandings" of the law. Treasury agents and federal prosecutors dropped most of the cases they received, in part because they had little or no idea what the boundaries of the law really were. Action was still pending in 699 cases, but only 106 convictions had been obtained.

Individual doctors were also confused. Many wrote to the government to urge that they be allowed to continue prescribing morphine to longtime addict patients. One Tennessee doctor pleaded with President Woodrow Wilson to allow for a way "to help the unfortunate . . . I know what the suffering will be."[46] Even after two years, Alfred Gordon asked the New York Medico-Legal Society in 1917: "can any one say that the desired results have been obtained?" For Gordon, two years had still brought no clear answer to the key questions:

> Have the authors of the Harrison Law considered the untold suffering inflicted on the victims of the disease called "Morphinism" by suddenly withdrawing the drug from them? Has the new law provided for the great army of sufferers that are left on our hands because of their helplessness due to compulsory abstinence? Has it actually placed in the power of the country a prophylactic measure to prevent moral and intellectual degeneration, which, indeed, should be its chief aim?[47]

Anticipating the debates of decades to come, Gordon concluded, "medical science will continue its investigations into the underlying causes of these diseases and render its valuable assistance in formulating biological and therapeutic laws for an intelligent manipulation of the grave problem of drug addiction."[48]

Alone in their exclusion from the process of framing the Harrison Narcotics Act, drug users had their own response to the new law. For some, 1914 represented more of the same. Cocaine users had felt the brunt of the cocaine wars for years, and the so-called dope fiends had long been denied easy access to their drugs of choice, harassed by the police, jailed for vagrancy, and frustrated by the absence of medical assistance.

For the legitimate or respectable addict, the Harrison Narcotics Act created terror. For decades, opiate addiction among the respectable classes had been quietly tolerated, if rarely discussed. Now this tolerance seemed to be coming to an end. Just two weeks after the law went into effect, a letter addressed "To Anyone who has power to amend the new Harrison Drug Law" made its way to the U.S. Public Health Service. The letter, from a Mrs. P. A. of Chicago, is quoted in full to underline the dread anticipation felt in some quarters:

> I am writing this urgent appeal, as I have been advised by people who are powerless to help, although they feel like murderers for not doing so, to ask if there is not some way—an

> emergency law—could be passed at once by which old people could be allowed the drug that is alone life to them.
>
> My father had used morphine for nearly forty years—and has just finished his last, and we cannot obtain it for him, although we know he will die without it and the doctors tell us so too. The cure is no hope to us as he can not stand it at his age and having taken it so long. So in the next few days or a week, he is facing one of the most horrible deaths imaginable. It seems terrible to me that a law should be passed which will deliberately kill off hundreds of people.
>
> The law is one of the finest things that has ever come to our country, for the drug ruined our home from the very beginning, but now that my father is old, I could not take it away from him just to kill him—and it does not seem right that any one should.[49]

Hope, fear, and uncertainty would be answered by practice. By 1919 the federal government would show more clearly the direction of drug prohibition in the United States.

URLS FOR PRIMARY DOCUMENTS

Pure Food and Drug Act of 1906
<http://coursesa.matrix.msu.edu/~hst203/documents/pure.html>

"Negro Cocaine 'Fiends' New Southern Menace," February 8, 1913, *The New York Times*
<http://www.druglibrary.org/schaffer/History/negro_cocaine_fiends.htm>

"Say Drug Habit Grips Nation," December 5, 1913, *The New York Times*
<http://www.druglibrary.org/schaffer/history/e1910/say_drug_ habit.htm>

"New Cocaine Bill Adds to Penalties," January 17, 1913, *The New York Times*
<http://www.druglibrary.org/schaffer/history/e1910/newcocainebill.htm>

Harrison Narcotics Tax Act, 1914
<http://www.druglibrary.org/schaffer/history/e1910/harrisonact.htm>

The Indian Hemp Drugs Commission Report
<http://www.drugtext.org/library/reports/inhemp/ihmenu.htm>

"Interpret Harrison Law Supreme Court Decides It Only Applies to Those Who Deal in Drugs," June 6, 1916, *The New York Times* <http://www.druglibrary.org/schaffer/history/e1910/interpret_harrison_act.htm>

Legal References on Drug Policy Federal Court Decisions on Drugs by Decade 1910
<http://www.druglibrary.org/schaffer/legal/legal1910.htm>

NOTES

1. Harrison Narcotics Act, PL 63-223, 38 Stat. 785; amended February 24, 1919, by PL 65-254, 40 Stat. 1057, 1130.

2. On the passage of the Harrison Act and legislative developments, see David F. Musto, *The American Disease: Origins of Narcotic Control,* Third Edition (New York: Oxford University Press, 1999).

3. The development of the American ethical pharmaceutical industry has been the subject of too few historical studies, but this small number includes some extremely valuable works, among them: Louis Galambos with Jane Elliot Sewell, *Networks of Innovation: Vaccine Development at Merck, Sharp and Dohme, and Mulford, 1895-1995* (New York: Cambridge University Press, 1995); Jonathan Liebenau, *Medical Science and Medical Industry: The Formation of the American Pharmaceutical Industry* (Baltimore: The Johns Hopkins University Press, 1987); John Parascandola, *The Development of American Pharmacology: John J. Abel and the Shaping of a Discipline* (Baltimore: Johns Hopkins University Press, 1992); and John P. Swann, *Academic Scientists and the Pharmaceutical Industry: Cooperative Research in Twentieth-Century America* (Baltimore: The Johns Hopkins University Press, 1988). On the Parke, Davis research, see the text on pharmaceutical chemistry published by the company's first chemist: Albert B. Lyons, *A Manual of Practical Pharmaceutical Assaying* (Detroit: D. O. Haynes and Co., 1886).

4. F. E. Stewart, "Brief History of the Founding of the Scientific Department of Parke, Davis and Co." (unpublished manuscript, n.d.), Francis Edward Stewart Papers, MSS 606, Archives Division, State Historical Society of Wisconsin, American Institute of the History of Pharmacy Collection; see also Interview with F. W. Robinson, 1958, pp. 10-11. MS, Parke, Davis and Company Collection, Burton Historical Collection, Detroit Public Library.

5. F. E. Stewart, "Coca-Leaf Cigars and Cigarettes," *Philadelphia Medical Times* 15 (September 19, 1885): 935.

6. Lewis D. Mason, "Patent and Proprietary Medicines As the Cause of the Alcohol and Opium Habit or Other Forms of Narcomania—with Some Suggestions As

to How the Evil May Be Remedied," *Quarterly Journal of Inebriety* 25 (January 1903): 6.

7. James Harvey Young is a leading historian of patent medicines whose long career has detailed the persistent strain of health fraud and quackery in American history. Young, *American Health Quackery: Collected Essays by James Harvey Young* (Princeton, NJ: Princeton University Press, 1992), summarizes much of this work. The work of J. Worth Estes provides an interesting counterpoint to Young's. See particularly J. Worth Estes, "The Pharmacology of Nineteenth Century Patent Medicines," *Pharmacy in History* 30 (1988): 3-18.

8. See, for example, Charles N. Crittenton, *Catalogue of Proprietary Medicines and Druggists' Sundries* (New York: Charles N. Crittenton, 1891).

9. "The Druggist As a Citizen," *American Druggist and Pharmaceutical Record* 61 (October 1913): 22.

10. "Cocaine Legislation," *Midland Druggist* 4 (1903): 713.

11. Joseph F. Spillane, *Cocaine: From Medical Marvel to Modern Menace in the United States, 1884-1920* (Baltimore: The Johns Hopkins University Press, 2000), Chapters 1 and 2.

12. The policy literature on medicalization is quite large. See Avram Goldstein and Harold Kalant, "Drug Policy: Striking the Right Balance," *Science* 249 (1990): 1513-1522. See also Don C. DesJarlais, "Harm Reduction: A Framework for Incorporating Science into Drug Policy," *American Journal of Public Health* 85 (1995): 10-12, in which he observes that "the harm reduction perspective emphasizes the need to base policy on research rather than on stereotypes of legal and illegal drug users."

13. J. B. Woodhull, "The Drug Habit and Legislation," *New York Medical Journal* 101 (January 2, 1915): 20.

14. Ibid., p. 20.

15. "Public Waking up to Cocaine Menace," *The New York Times,* August 3, 1908, Section 5, p. 5.

16. Ibid.

17. *Limiting Production of Habit-Forming Drugs and Raw Materials from Which They Are Made: Hearings Before the Committee on Foreign Affairs,* 67th Congress, 4th Session, 1923, pp. 55-56.

18. Heine Marks, "The General Treatment of Habitual and Periodic Alcoholic, Morphine, and Cocaine Inebriates," *Quarterly Journal of Inebriety* 18 (1896): 159.

19. In reading these early medical studies, one cannot help but be struck by the sophistication of physicians' treatment of cocaine, and their attentiveness to the effects of form, dosage, route of administration, and even setting.

20. E. R. Waterhouse, "Cocaine Debauchery," *Eclectic Medical Journal* 56 (1896): 464.

21. Edward Huntington Williams, "The Drug Habit Menace in the South," *Medical Record* 85 (1914): 247-249.

22. Charles A. Bunting, *Hope for the Victims of Alcohol, Opium, Morphine, Cocaine and Other Vices* (New York: Christian Home Building, 1888), p. 71.

23. Foster Kennedy, *New York Medical Journal* 11 (1914): 20-22.

24. For the most complete review of the existing evidence on the size of the American addict population, see David T. Courtwright, *Dark Paradise* (Cambridge, MA: Harvard University Press, 1982).

25. "New Danger from the Use of Heroin," *American Druggist and Pharmaceutical Record* 61 (May 1913): 20-21.

26. James H. Beal, *An Anti-Narcotic Law* (Detroit: William M. Warren, 1903), p. 1.

27. Rufus King, "The Narcotics Bureau and the Harrison Act: Jailing the Healers and the Sick," *Yale Law Journal* 62 (1953): 784. The pattern of early state and local activity holds true for the 1937 Marihuana Tax Act as well. Richard J. Bonnie and Charles H. Whitebread II noted that, by 1937, no fewer than twenty-seven states had passed marijuana laws. See Bonnie and Whitebread, "The Forbidden Fruit and the Tree of Knowledge: An Inquiry into the Legal History of American Marijuana Prohibition," *Virginia Law Review* 56 (October 1970): 971.

28. Jeffrey Clayton Foster, "The Rocky Road to a 'Drug Free Tennessee': A History of the Early Regulation of Cocaine and the Opiates, 1897-1913," *Journal of Social History* 29 (spring 1996): 547-564.

29. Ronald Hamowy (Ed.), *Dealing with Drugs: Consequences of Government Control* (Lexington, MA: Lexington Books, 1987).

30. Martin I. Wilbert and Murray Galt Motter, *Digest of Laws and Regulations in Force in the United States Relating to the Possession, Use, Sale, and Manufacture of Poisons and Habit-Forming Drugs*, Public Health Bulletin no. 56, November (Washington, DC: Government Printing Office, 1912).

31. Spillane, *Cocaine,* pp. 147-148.

32. Hamowy, *Dealing with Drugs,* p. 13.

33. W. J. O'Connor, inspector of police to Dr. Hamilton Wright (June 22, 1909), Hamilton Wright Papers, National Archives; "New York to Have Drastic Cocaine Law," *Pharmaceutical Era* (February 1913): 96.

34. David Musto, "The History of Legislative Control over Opium, Cocaine, and Their Derivatives," in Ronald Hamowy (Ed.), *Dealing with Drugs: Consequences of Government Control* (Lexington, MA: Lexington Books, 1987), p. 46.

35. On Tuscon, see Neil B. Carmody (Ed.), *Whiskey, Six-Guns & Red-Light Ladies: George Hand's Saloon Diary, Tucson, 1875-1878* (Silver City, NM: High-Lonesome Books, 1994), p. 248; Roger McGrath, *Gunfighters, Highwaymen, and Vigilantes: Violence on the Frontier* (Berkeley: University of California Press, 1984), pp. 124-130. On the Montana newspapers, see for example "Chinatown Explosion," *Butte Miner,* November 11, 1914; "Hop Sing Running Opium Den," *Butte Miner,* August 20, 1903; "Chinese Hop Joints Are Raided," *Butte Miner,* June 26, 1905.

36. Chicago Department of Police, *Annual Report of the General Superintendent of Police of the City of Chicago,* 1902.

37. Joseph Spillane, "The Making of an Underground Market: Drug Selling in Chicago, 1900-1940," *Journal of Social History* 32 (fall 1998): 27-47.

38. See Sarah W. Tracy, "The Foxborough Experiment: Medicalizing Inebriety at the Massachusetts Hospital for Dipsomaniacs and Inebriates, 1833-1919" (doctoral dissertation, University of Pennsylvania, 1992) and Jim Baumohl and Sarah W. Tracy, "Building Systems to Manage Inebriates: The Divergent Paths of California and Massachusetts, 1891-1920," unpublished paper.

39. Samuel Hopkins Adams, *The Great American Fraud* (Chicago: American Medical Association, 1913). For a still useful and entertaining narrative of Adams'

role in helping to secure passage of the Food and Drug Act of 1906, see Mark Sullivan, *Our Times* (New York: Charles Scribner's Sons, 1929).

40. James Harvey Young, *Pure Food: Securing the Federal Food and Drugs Act of 1906* (Princeton, NJ: Princeton University Press, 1989).

41. Spillane, *Cocaine,* p. 131.

42. Musto, *The American Disease* offers a good summary of these early developments. A fuller treatment may be found in William B. McAllister, *Drug Diplomacy in the Twentieth Century: An International History* (New York: Routledge, 2000).

43. See Anne Cipriano Venzon, "Francis Burton Harrison," in *American National Biography,* Volume 10 (New York: Oxford University Press, 1999), pp. 208-209. Harrison, even more than most Democrats at the time, had been a critic of U.S. policy in the Philippines and an advocate for independence. In the eight years Harrison served as governor-general, he worked toward a liberalization of Philippine policy, work that was to some extent undone by subsequent Republican administrations.

44. "Federal Aid in the AntiDrug War," *The Literary Digest* (April 24, 1915): 958.

45. "Habit Forming Drugs," *Outlook* (May 5, 1915): 8-9. The *New York Herald*'s observations were cited in this article also.

46. This letter is from File 2123, RG 90, Records of the United States Public Health Service, National Archives, Washington, DC.

47. Alfred Gordon, "The Relation of Legislative Acts to the Problem of Drug Addiction," *Journal of Criminal Law and Criminology* (July 1917): 211-215. Gordon's concerns for the addict represented one set of questions about the Harrison Act. Those who felt the act to be too lenient and full of loopholes, on the other hand, looked toward the day when the law would "deal more effectively [not only] with the petty, illicit dealer, but also to close in upon the medical practitioners who, under cover of a gradual reduction treatment have been competing with the peddler of the tenderloin in enabling addicts to get their drugs." Francis Fisher Kane, "Drugs and Crime," *Journal of Criminal Law and Criminology* (November 1911): 357.

48. Gordon, "The Relation of Legislative Acts," p. 215.

49. This letter is from File 2123, RG 90, Records of the United States Public Health Service, National Archives, Washington, DC.

Chapter 2

Building a Drug Control Regime, 1919-1930

Joseph F. Spillane

Passage of the Harrison Narcotics Act did not create a coherent drug policy. The medical profession continued its debates over the nature of addiction and its appropriate treatment. Local governments endorsed a bewildering variety of approaches to their own drug problems, from strict enforcement of punitive criminal laws to addict maintenance at narcotic clinics. Influential policy voices within the federal government divided over the merits of a public health-based policy versus an approach favoring criminal law and law enforcement. In addition, the Supreme Court had shaken the entire legal foundation of the Harrison Narcotics Act with its decision in *United States v. Jin Fuey Moy* (1916). In that case, the Court had refused to endorse the government's position that the possession of narcotic drugs constituted presumptive evidence of a violation, thus limiting federal police powers in drug enforcement.[1]

One might have been forgiven, at the start of 1919, for assuming that the muddled state of drug policy would be an ongoing concern. Instead, 1919 proved to be the year in which the basic outlines of federal drug policy came into focus. That year, the Treasury Department and the Public Health Service laid the foundations for a unified, official story of drugs and addiction that would dominate the next several decades. In March 1919, a divided Supreme Court endorsed the government's stricter interpretation of the Harrison Narcotics Act. At the end of the year, responsibility for the act's enforcement shifted to the new Narcotic Division of the Treasury Department's Prohibition Unit. Under the leadership of Levi Nutt, the Narcotic Division would prove instrumental in defining drug abuse as a police problem.

CHARTING AN OFFICIAL POLICY

The year 1919 seems an appropriate moment for federal drug prohibition to assert itself. World War I, recently concluded, left a legacy of national concern for shielding the country from "foreign" dangers and from internal weakness and deviance. The recent Espionage Act (1917) and Sedition Act (1918) strengthened the power of the federal government to confront perceived threats to national order and domestic security. A campaign to crush political radicalism and labor activism expanded federal police activity, marked notably by the creation of a General Intelligence Division within the Justice Department's Bureau of Investigation and the "Palmer" raids, launched in November 1919. The latter featured the arrest and detention of over 4,000 men and women on suspicion of violating the wartime Sedition Act. That same year, the Eighteenth Amendment was added to the U.S. Constitution, marking the formal starting point of national alcohol prohibition. In this climate, an official drug policy emerged.[2]

THE RETREAT OF THE TREATMENT AND MAINTENANCE MODELS

In the summer of 1919, Senator Joseph I. France, one of the few physicians in the Senate, introduced a bill that would have authorized the use of Public Health Service hospitals for the treatment of drug addicts, as well as federal matching funds for publicly funded drug treatment programs at the state and local levels. The France Bill, as it came to be known, was actually put together by the commissioner of internal revenue, Daniel C. Roper. He was an experienced administrator who quickly concluded that addiction was not readily solved as a police problem. Instead, Roper believed that the drug problem would be better defined as a medical issue requiring an investment of treatment resources. That same summer Roper had authored an internal memo to internal revenue agents charged with enforcing the Harrison Narcotics Act, directing them to assist local authorities in aiding addicts. Roper's memo implicitly endorsed the maintenance of the drug addict; numerous municipalities established narcotic clinics under what they now assumed was the protection of the federal government.[3]

The summer of 1919 proved to be the high-water mark for a public health-centered approach to America's drug problem. The France Bill came out of committee, but died in the full Senate in the fall. The 1918 congressional elections had returned a Republican majority to the Senate, one that was not inclined to support the ambitious federal treatment plan. By the end of 1919, the Treasury Department had essentially repudiated Roper's own position favoring the maintenance of addicts. More than forty municipal drug clinics, many just opened, now found themselves at odds with federal drug policy. Federal pressure resulted in the closure of most of these clinics by 1921. The last public clinic closed in Shreveport, Louisiana, in early 1923.

The sudden failure of the public health model reflected a confluence of unfavorable developments, most of which came from within the medical profession itself. The increasingly conservative leadership of the AMA, who would express outrage over the passage of the 1921 Sheppard-Towner Act providing federal funding for maternal and child health care, viewed the France Bill as an equally suspect example of government medicine. More important, medical opinion that had formerly supported treatment and maintenance shifted in ways that undercut both of these policy approaches.

Clear evidence of this changing medical opinion was expressed in a memorandum from A. G. DuMez to the surgeon general in February 1919. DuMez was the Public Health Service's primary narcotic "expert," having replaced the late Martin Wilbert the previous year. DuMez had also been appointed to the Special Narcotic Committee of the Treasury Department. His membership on this committee, charged with rethinking the nation's approach to its drug problems, placed DuMez in an especially important position. In his memorandum, DuMez reported the new conventional wisdom in the medical field—that addiction cures failed to provide permanent help and that most addicts continued their habits despite determined intervention. "Our present methods of treating drug addiction must be considered failures," he concluded. This dismal report would reinforce the inclination of the Public Health Service to withhold its support from the France Bill and any substantial commitment to provide treatment.[4]

The Public Health Service also led the fight against the so-called antibody theory of addiction. This theory had, until World War I, been a fairly popular medical approach to understanding the addiction phenomenon. The theory hypothesized that addicts developed specific antibodies in the blood that prevented the discontinuation of drug use. These physiological changes were beyond the control of the addict, and many doctors accepted that these changes required maintenance doses to be given indefinitely. Many of the leading supporters of the narcotic clinics had been schooled in versions of the antibody theory, including such influential narcotics experts as Dr. Ernest Bishop in New York and Dr. Edward Huntington Williams in Los Angeles.[5]

No scientific evidence existed by 1919 to support the antibody theory, nor was any forthcoming, and a new medical orthodoxy embraced a different view. The newer theory emphasized the idea that most addicts had some "form of mental defect, psychopathic personality, or mental impairment from drugs, which in terms of will power meant impaired ability to resist."[6] The Public Health Service came to endorse this revised explanation for narcotic use. In 1925 Dr. Lawrence Kolb of the Public Health Service authored a series of articles in *Mental Hygiene* that described addicts, or at least those who chose their addiction, as psychopathic personalities.[7] For Kolb, the best that could be done for such people was to try and keep them away from drugs. This, he felt, was primarily a function of law enforcement. Kolb's influence may be seen in the remarks of Public Health Service chief H. S. Cumming in 1925. "The vast majority of those who have become addicted in recent years," Cumming observed, "have deliberately sought narcotics and they use drugs as a form of dissipation." Further, these addicts "are recruited from among a neurotic or psychopathic type of persons, many of whom would have been police problems even if they had never become addicted."[8] Thus the nation's leading medical experts on addiction were effusive in their praise for strict enforcement of the Harrison Narcotics Act, observing as Kolb and DuMez did that it was a great weapon in "reducing the extent of addiction."[9]

The AMA quickly accepted the antimaintenance view. Early in 1920 the AMA's Committee on Habit-Forming Drugs had adopted a policy of opposition to ambulatory maintenance clinics, a policy reaffirmed by the AMA's House of Delegates at the annual meeting in April 1920. Alfred C. Prentice, a member of the AMA committee, criticized the "shallow pretense that addiction is a disease."[10] The following year, the AMA's Council on Health and Public Instruction would again reaffirm the organization's antimaintenance position.

The AMA's stand did not foreclose continued debates among individual physicians. In 1920 Dr. Thomas Blair condemned the "easy" doctor who "can't say no" to patients seeking to maintain addictions, and asserted his view that the Harrison Narcotics Act did nothing more than what would be recommended in "the standard literature of the professions." Blair's position provoked a response from one rural Pennsylvania physician, who defended his practice of prescribing opiates in vivid terms:

> I wonder if you can conceive of how hard the backwoods women work and how wholly devoid of amusement or pleasure their lives are? They do not even have comfortable beds or chairs; for the most part, no bathing facilities but the washtub. . . . For the most part they live their lives as you might have expected pioneers of seventy-five years ago, but without the game to be had for the shooting. They are seven miles from a railroad station, although the log railway carries a freight car now and then besides lumber cars . . . and the ________ Lumber Company strongly forced down production costs regardless of the health of employees.[11]

After receiving the letter, even Dr. Blair stated, "it is not fair to call the physician so placed a dope-seller. What is he to do?"[12] Honest debate within the profession faded after 1919, however, and the minority of physicians who sought to maintain addicts, or even employ gradual reduction treatments, lost whatever professional protection they may once have enjoyed.

In October 1919, Congress passed the Volstead Act, legislation that provided the operational details for the enforcement of the Eighteenth Amendment prohibiting alcohol. The Volstead Act authorized the creation of a Prohibition Unit within the Treasury Department's Bureau of Internal Revenue. The Prohibition Unit included the Narcotic Division, the first agency specifically organized around enforcement of the Harrison Narcotics Act.[13] The division, headed by Levi Nutt, was given nearly double the money spent on drug law enforcement the previous year, and received authorization for 170 narcotics agents to work in district offices nationwide.

During the winter of 1919, Nutt began the task of reshaping federal enforcement policy. With the pro-maintenance Roper now out of the direct policy process, Nutt determined that the Narcotic Division would take a firmly antimaintenance stance. The cornerstone of that policy would be a vigorous assault on the municipal narcotics clinics. Roper's revenue agents had cooperated with, or at least tolerated, the clinics. Nutt's narcotic agents, in contrast, had a singular focus on closing these programs.[14]

Although there was considerable variation among the clinics, most shared at least some features of the Newark, New Jersey, clinic. Every Monday, Wednesday, and Friday the clinic opened its doors at 3:00 p.m. Only morphine was available, not cocaine or heroin, and it was distributed to addicts with a view toward achieving a gradual reduction cure. The doctors who ran the clinic relied on the antitoxin theory, or what they called "the chronic toxemia" that the addict suffered. Curing addiction was probably no more than a secondary goal to dealing with the toxemia.[15]

A review of Louisiana clinics in *The American City* magazine listed the guiding principles of the clinic model:

> We realize that a permanent cure of those afflicted with drug-addiction disease is impossible, in the great majority of cases, unless the addict be placed in a position to secure scientific treatment. The sole object of this dispensary is to relieve suffering until such time as a scientific treatment may be had.
>
> The basis of operation is legitimate supply versus illegitimate trafficking. To prevent a victimized people from being more thoroughly victimized by heartless, profiteering ghouls. To prevent the making of new addicts.
>
> Diminishing petty thievery, which constitutes a tax or burden on society, for the reason that many addicts, unable to pay the price of from $1 to $3 a grain, are forced to criminal methods.[16]

Statement of principles aside, there was little the clinics could do in the face of the combined opposition of the medical profession and the Narcotics Division. In effect, the clinics found themselves without any institutional support that might have allowed them to survive. Rather than being incorporated into larger public health systems, maintenance providers most often found themselves isolated. Anticlinic operations began in mid-1920 and, within a year, virtually all of the maintenance centers had either closed or were in the process of shutting down. Over the next four decades, government policy would consistently endorse the same basic principles: the best treatment for the intractable condition of addiction involved separating the addict from the supply; ambulatory maintenance was both illegitimate and a real-world failure; and law enforcement must occupy the central position in antidrug efforts.[17]

OFFICIAL CONTROL OF INFORMATION

Despite the fact that Colonel Nutt would have preferred to simply dictate policy without regard to popular or political opinion, the position of the federal drug control apparatus was far too precarious. Survival of this government agency meant justifying the existence of a war on drugs, for which the control and dissemination of information to the public was critical. Nowhere is this better illustrated than in the case of the "one million United States addicts."

In June 1919, the Special Narcotic Committee of the Treasury Department issued its final report, *Traffic in Narcotic Drugs.*[18] The report reflected a year's worth of work by a committee whose members were widely regarded as the leading experts in the field. While the report covered many aspects of the drug problem, no part of the report drew more attention than its estimate of the nation's addict population. The committee had surveyed physicians and pharmacists registered under the Harrison Narcotics Act, asking them to report the number of addicts under treatment. Only 30 percent of doctors replied, and these reports identified 73,000 addicts under treatment. The committee extrapolated from this figure that the actual number of addicts would be 238,000 if all the physicians had replied. Then, in an even bigger statistical leap of faith, the committee estimated that since not all addicts were receiving treatment, the real number of addicts was 1 million. The figure of 1 million seems to have come directly from guesswork and a rough compromise between a figure of 750,000 generated by the Public Health Service and 1.5 million generated by the Internal Revenue Service.[19]

Coming as it did at a time of considerable public concern over the drug problem, the figure of 1 million was widely reported, almost invariably as the shocking and "true" portrait of the American drug problem.[20] What is less well known, however, is the active role the committee played in disseminating the figure of 1 million to journalists. Well before the report appeared, even before completion of the study, committee sources passed along previews for news outlets. *The New York Times* ran a story titled "A Million Drug Fiends" on September 13, 1918.[21] The source for much of this misinformation seems to have been the office of committee chairman Henry T. Rainey, Representative from Illinois and future Speaker of the House.

Those close to the drug problem reacted with deep skepticism. It was reported "the Pennsylvania Department of Health, which registers all kinds of addicts as required by its own law, cannot begin to find one hundred thousand addicts within the state and does not believe there are nearly that number."[22] Pearce Bailey, formerly the Army's chief neuropsychiatrist and no friend of the addict (he described them as "the tools of designing propagandists in spreading seditious doctrines, or in the commission of acts in defiance of law and order") harshly criticized the report in *The New Republic*.[23] Subsequent studies have confirmed that these critics were right. There were nowhere near 1 million addicts in the United States in 1919, and probably not even one-fourth of that number.[24] In the overheated atmosphere of 1919 there was little opportunity for critical analysis of the Treasury report. Informed critiques received scant attention, while federal enforcement policy received a critical boost.

Within a few years, the government effectively repudiated the idea that there were, or had been, 1 million addicts in the United States. With the narcotic clinics closed and the Narcotic Division placed on a firm footing, the political imperative shifted from mobilizing public opinion to demonstrating effective policies. In 1924 Lawrence Kolb and A. G. DuMez estimated the nation's addict population at about 110,000.[25] This study, far more accurate than the 1919 Treasury estimate, proved valuable politically. Colonel Nutt used the figure of 110,000 the following year in testimony before Congress, and downgraded that figure to 95,000 in 1926 testimony before the House Appropriation Committee.

The "1 million addicts" figure, however, refused to go away. Journalists continued to use the number in stories; the *Literary Digest* cited the figure of 1 million as a "conservative estimate" in 1923.[26] Still worse, from the government's perspective, the number became a weapon that critics of government inaction could use to stir public indignation. The most resourceful of these critics was Richmond Pearson Hobson. The quintessential "moral entrepreneur," Hobson had begun his career attacking alcohol before attaching himself to the drug problem in the mid-1920s.

For a time it seemed that the control of the official "story" of drugs would pass from the Public Health Service and the Narcotic Division to Hobson. His organization, the International Narcotic Education Association, established in 1923, worked tirelessly to alert the American public and, especially, wealthy potential donors to the menace and peril of narcotic drugs. In *The New York Times* of November 9, 1924, Hobson was the feature source for the story "One Million Americans Victims of Drug Habit" in which Hobson also argued "crime gallops with heroin." That same year, Hobson very nearly persuaded Congress to reprint 50 million copies of his pamphlet "The Peril of Narcotic Drugs" as a public service. It was eventually placed in the *Congressional Record* instead.[27] His powerful allies in Congress included Senator Royal S. Copeland, former New York City health commissioner and strong advocate of antidrug policies. In 1927 Copeland introduced a Hobson-generated bill for the establishment of a U.S. college of narcotic education, "with a student body to consist of students appointed two by each Senator, two by each Member of the House of Representatives, and twenty by the President."[28]

Hobson had powerful media connections, with radio networks donating time for speeches and newspapers publishing Hobson-authored scare stories. Hobson's most influential ally was William Randolph Hearst, whose newspapers provided the most consistent forum for antidrug publicity before World War II. Hearst papers were never shy about pushing the drug issue. In the summer of 1925, Hearst was backing New York Mayor John Hylan for reelection against Jimmy Walker; one cartoon showed Walker looking on approvingly as a drug peddler sold drugs to a child! An editor of Hearst's Chicago *Herald-Examiner* confessed that the drug issue was just one of many that could periodically be packaged as a "crisis":

> we just do what the Old Man orders. One week he orders a campaign against rats. The next week he orders a campaign against dope peddlers. Pretty soon he's going to campaign against college professors. It's all the bunk, but orders are orders.[29]

In his campaign Hobson played a dangerous game, simultaneously doing all he could to give the appearance of government support, while at the same using government inaction as one of his favorite negative themes. The critical showdown began in 1926 in the midst of preparations for a World Narcotic Conference for which Hobson hoped to secure government endorsement. The Public Health Service sought to head off the conference or at least deprive Hobson of formal support from any federal officials. Lawrence Kolb asserted that

> the executive departments having to do with narcotic education or with the Philadelphia exposition look upon his conference with contempt and are resentful because of the methods he has used to spread the impression that these departments support his movement. . . . there is also resentment because of the persistent and unscrupulous efforts he has made to bring about the endorsement of his general program by government agencies that he has been told can never honestly endorse it.[30]

Surgeon General Cumming opposed Hobson's campaign, arguing that "narcotic addiction is a real problem, but one which is successfully being dealt with." Colonel Nutt, for his part, wrote of Hobson's drug pamphlet: "it was, therefore, our duty both in the interest of truth and public health, to give information that would prevent such a document from being endorsed and broadcasted by the Government."[31]

The government had learned from experience.[32] A decade later a drug agent informed Indiana University sociologist and Anslinger critic Alfred Lindesmith that his agency had a law enforcement duty but also the duty of "disseminating right information and preventing the dissemination of wrong information." Lindesmith later reflected that this could work against any critic of government policy, including both the critics of law enforcement-centered approaches to the drug problem (including Lindesmith himself) and those who sought to overdramatize the problem and condemn the government for inaction.[33]

DRUGS, THE COURTS, AND THE FEDERAL CRIMINAL LAW

The Harrison Narcotics Act arrived at a time when federal criminal law was just beginning to assert itself. For much of the nation's history, federal police power had been limited to a few areas, such as counterfeiting, piracy, and crimes committed on federal lands. Additions to the federal criminal law tended to address only those questions where state law was inadequate, such as the Post Office Act of 1872, used to attack the use of the mail to conduct lotteries and to perpetrate frauds.

The expansion of federal regulatory power at the end of the nineteenth century, including the Interstate Commerce Act (1887) and the Sherman Anti-Trust Act (1890), opened the door to further federal criminal law through the use of congressional power to regulate interstate commerce. A series of federal criminal laws took their authority from the commerce clause, including the Mann Act (1910), the so-called white slave act prohibiting the transportation of a woman across state lines for immoral purposes, and the Dyer Act (1919), prohibiting the transportation of a stolen motor vehicle across state lines.[34] The Supreme Court's endorsement of both of these acts tended to emphasize the inability of individual states to effectively deal with these problems because of changes caused by modern transportation and communication. The reasoning, then, was pragmatic. In *Brooks v. United States* (1925), upholding the Dyer Act, Chief Justice Taft urged the necessity of the federal government stopping "evil minded persons" where states could not.[35]

In spite of its endorsement of certain extensions of federal power, the Supreme Court was not inclined to blindly support federal intrusions into what it viewed as the legitimate province of the states. Indeed, the Court in this era was deeply divided between a core of conservative justices who urged a narrow reading of federal power and a "progressive" group who sought to allow the national government to assert itself in the cause of the public welfare. In *Hammer v. Dagenhart* (1918), the Court refused to sanction a federal statute that had prohibited transportation in interstate commerce of goods made in factories that employed children. Though the statute did not in specific terms interfere with local production or manufacturing, there was little question that its real purpose was to suppress child labor on a national level. The majority found that this primary purpose rendered the law invalid.[36]

Opponents of child labor, having failed through the exercise of the federal government's power to regulate interstate commerce, next turned to the power to tax. After the *Hammer v. Dagenhart* setback, Congress passed a 10 percent tax upon the profits of all persons employing children. Once again, the Supreme Court ruled that since the purpose of the tax measure was "actually" to eliminate child labor, this primary motive invalidated the tax. Chief Justice Taft, writing for the majority in what came to be known as the *Child Labor Tax Case* (1922), expressed his view that the desire to end child labor was undoubtedly the desire of "good people." "Unfortunately," Taft wrote, "we cannot strain the Constitution to meet the wishes of good people."[37]

The Harrison Narcotics Act presented the Court with similar questions. Was the control of narcotics an area that should, or could, be effectively dealt with by the individual states? Could the primary purpose of the Harrison Narcotics Act fairly be said to be the imposition of a tax, or was this purpose merely secondary to the reduction of drug use in the United States?

The Court first dealt with these questions in *United States v. Jin Fuey Moy* (1916). The Court quickly condemned Section 8 of the original act that made possession of narcotics presumptive evidence of a violation. Since the Harrison Narcotics Act required registration of only those who "import, produce, manufacture, deal in, dispense, sell or distribute" narcotic drugs, no one not required to register could be punished merely for having drugs in their possession. Further, the Court expressed doubts that physicians could be prevented from prescribing drugs to addicts, and stopped just short of questioning whether the entire act was unconstitutional.[38]

The Court confronted the question of the act's constitutionality directly in *U.S. v. Doremus* (1919). *Doremus* provided the first in a line of decisions affirming the primacy of revenue collection to the Harrison Narcotics Act. The opinion was written by Justice William R. Day, a long-serving but largely obscure member of the Court.[39] Day asserted "the Act may not be declared unconstitutional because its effect may be to accomplish another purpose as well as the raising of revenue. If the legislation is within the taxing authority of Congress—that is sufficient to sustain it." Day asked the question: "Have the provisions in question [those that limited sales to registered dealers and physician prescription] any relation to the raising of revenue?" He felt that they clearly did, writing,

> they tend to keep the traffic aboveboard and subject to inspection by those authorized to collect the revenue. They tend to diminish the opportunity of unauthorized persons to obtain the

> drugs and sell them clandestinely without paying the tax imposed by the federal law.[40]

Significantly, Day concluded that no purpose other than the facilitation of tax collection would invalidate the Harrison Narcotics Act.

Day's interpretation of the basic function of the act was never successfully challenged. Of course, the Harrison Narcotics Act did generate significant revenue for the federal government, particularly after the act was amended in 1918 to require a tax stamp to be placed on every package. Federal enforcement activity, on the other hand, seems to show what the Court denied—that the Harrison Narcotics Act was intended to regulate medical practice and eliminate certain kinds of drug use, not merely raise revenue. Nonetheless, the Court repeatedly affirmed this core holding of *Doremus* over the next decade.[41] In 1928 Justice Holmes told critics that "it is too late to attempt to overthrow the whole act on Child Labor Tax Case . . . the statute is much more obviously a revenue measure now than when *United States v. Doremus* was decided, and is said to produce a considerable return."[42] Chief Justice Taft endorsed this same view the same year, in *Nigro v. United States,* pointing out that the Harrison Narcotics Act "has benefited the Treasury to the extent of nearly nine million dollars." If there was ever a doubt as to the character of the act, Taft wrote, "it has been removed . . ."[43]

Doremus answered only part of the question: Did the tax feature of the Harrison Narcotics Act pass constitutional muster? More problematic was the question of whether the act could legitimately dictate the boundaries of appropriate distribution and, by implication, use. *Webb et al. v. United States,* decided with *Doremus* in 1919, attempted to provide an answer. *Doremus* focused on Section 1 of the act, which required tax registration and limited that registration to certain groups. *Webb,* on the other hand, confronted Section 2 of the act, exempting from the order form requirements "the dispensing or distribution of any of the aforesaid drugs to a patient by a physician, dentist, or veterinary surgeon regularly registered under this act in the course of his professional practice only." The language—"in the course of professional practice only"—clearly implied some attempt to define the boundaries of legitimate medical practice. Could federal police power be extended to the regulation of medical practice?

In *Webb,* the Court gave only a partial answer. The Court's opinion, again authored by Justice Day, framed the question in this fashion: if a physician prescribed morphine to a "habitual user" for the purpose of maintaining that habit, without any attempt to cure the habit, was such action a "prescription" and thereby exempt from the requirements of the act? In this instance, the answer was not terribly clear. The defendant in the case was a "dope doctor" who had been peddling large quantities of morphine, selling prescriptions to anyone for fifty cents each—4,000 times in the months before his arrest. Day's opinion was short and to the point: to call "such an order for the use of morphine a physician's prescription would be so plain a perversion of meaning that no discussion of the subject is required."[44]

The central question in *Webb* was the power of the federal government to prohibit opiate maintenance by individual physicians, but the result did not settle the question. In the first place, no evidence showed that Webb attempted to justify his sales as "legitimate medical practice." Almost anyone might have agreed with Day that Webb's activities were obviously not within the boundaries of medical practice—but were they so far outside the boundaries as to avoid the question entirely? Moreover, the Court was deeply divided on both the *Doremus* and *Webb* cases, dividing 5-4 both times. Chief Justice White authored the *Doremus* dissent:

> The Chief Justice dissents because he is of the opinion that the court below correctly held the act of Congress, in so far as it embraced the matters complained of, to be beyond the constitutional power of Congress to enact because to such an extent the statute was a mere attempt by Congress to exert a power not delegated, that is, reserved police power of the states.[45]

The four dissenters (Chief Justices White, McKenna, Van Devanter, and McReynolds) were the most notably conservative justices on the court. That year the same four had dissented in another 5-4 split in the *Arizona Employers' Liability Cases,* in which they denied the constitutional validity of an Arizona workers' compensation law that McReynolds had found to be "a measure to stifle enterprise, produce discontent, strife, idleness, and pauperism."[46] Over time, resistance to *Doremus* faded, but the *Webb* precedent was anathema to the conservative justices.

The Court elaborated on the *Webb* decision in 1922, in *U.S. v. Behrman*. The case involved a New York physician who had been indicted for prescribing a large amount of heroin, morphine, and cocaine to one patient. While the amounts involved left little doubt that the facts represented less than legitimate medical practice, Behrman argued that it had been "good faith" prescribing.[47] Behrman urged "any construction which would forbid and penalize the giving of a prescription to afford temporary relief, even though a cure was not in immediate contemplation, would be a harsh construction not warranted by any language in the statute." The court, however, rejected the argument that anything said to be in "good faith" must be regarded as so. In light of the facts, said the Court, "not everything called a prescription is necessarily such."[48]

The *Behrman* case moved the Court a step closer to the Treasury Department's strict antimaintenance policy. In 1925 the Court was given the opportunity to adopt the government's position in its entirety. In *Linder v. U.S.*, a Spokane physician had been charged with a violation of the Harrison Narcotics Act after dispensing three tablets of cocaine and one tablet of morphine to an informer sent by federal agents. The jury charge in Dr. Linder's trial went further than *Webb* or *Behrman,* and formed the basis for appeal. The trial judge had instructed the jury that if the "defendant knew that this woman was addicted to the use of narcotics, and if he dispensed these drugs to her for the purpose of catering to her appetite or satisfying her cravings for the drug, he is guilty under the law."[49] In his defense, Dr. Linder asserted that his prescription constituted legitimate medical practice that the Harrison Narcotics Act had no power to touch.

A unanimous Supreme Court agreed with Linder, rejecting the federal government's antimaintenance position. Justice McReynolds, a dissenter in *Doremus* and *Webb,* wrote for the Court and directly challenged the core of the government's argument. In words nearly identical to those in the *Child Labor Tax Cases,* McReynolds declared that "federal power is delegated, and its prescribed limits must not be transcended even though the end seems desirable. . . . Congress never intended to interfere" with "professional conduct." *Behrman* could not be construed in the manner that the trial judge had done, because enforcement of the tax demanded no such "drastic rule."[50]

McReynolds went still further and attacked the very idea that addiction was an issue with which the police power of the state ought to be concerned. The Harrison Narcotics Act, he wrote, said nothing about "addicts" or their treatment. Addicts, McReynolds asserted,

> are diseased and proper subjects for such treatment, and we cannot possibly conclude that a physician acted improperly or unwisely or for other than medical purposes solely because he has dispensed to one of them in the ordinary course and in good faith, four small tablets of morphine or cocaine for relief of conditions incident to addiction. What constitutes bona fide medical practice must be determined upon consideration of evidence and attending circumstances.[51]

In *Boyd v. United States* (271 U.S. 104) the following year, the Court affirmed the *Linder* precedent. The government argued that doctors could prescribe only one dose at a time, a position the Court rejected. The *Linder* case threatened to undermine the very basis of government drug policy. Yet, years later, Alfred Lindesmith concluded "the *Linder* case had practically no effect and remains a ceremonial gesture of no practical significance for either addicts or physicians."[52]

Finally, the Supreme Court returned to the question of drug possession it had first dealt with in the *Jin Fuey Moy* case. The Treasury Department had responded to the Court's criticism in that case by persuading Congress to amend the Harrison Narcotics Act in 1918. Thereafter, all packages of narcotic drugs were required to bear a tax stamp—absence of this tax stamp would be evidence of illegal possession. For the Narcotic Division, the ability to prosecute possession was critical, given the relative difficulty in proving that an actual sale had taken place. In *United States v. Wong Sing* (1922) and *Yee Hem v. United States* (1925), the Supreme Court affirmed this position with regard to smoking opium.[53] This resulted in part because smoking opium had been banned in the United States entirely since 1909, so a presumption of illegal purchase seemed fairly simple.

The government's position was fully endorsed in 1928 with the decision in *Casey v. United States*.[54] The case involved a Seattle attorney who frequently represented drug defendants. A jailer had observed that some of Casey's clients appeared to be under the influence of narcotics after visits from their attorney. Based on this suspicion, the jailer induced one of Casey's clients to request morphine. When the morphine was delivered, Casey was arrested. In addition to arguing that he had been entrapped, the attorney made the case that no evidence of illegal purchase had been presented. The government argued again that the burden of proof was on the defendant to show that he had purchased the narcotics legitimately.

Writing for another 5-4 majority, Justice Oliver Wendell Holmes argued simply: "with regard to the presumption of the purchase of a thing manifestly not produced by the possessor, there is a rational connection between the fact proved and the ultimate fact presumed."[55] Predictably, Holmes' decision produced considerable skepticism among the conservative justices. Justice Butler offered the core of the dissent when he argued, "mere purchase or possession of morphine is not a crime. Congress has not attempted, and has no power, to make either an offense . . . there was no evidence to show how, when or where or from whom" Casey got the morphine. Butler concluded by suggesting that such prosecutions strained the idea of the Harrison Narcotics Act as a tax:

> It is hard to continue to say that this Act is a taxing measure in order to sustain it. Eagerness to use federal law as a police measure to combat the opium habit—a purpose for which Congress has no power to legislate—should not lead to the enactment or the construction of laws that shock common sense.[56]

Justice McReynolds went further:

> Once the thumbscrew and the following confession made conviction easy; but that method was crude and, I suppose, now would be declared unlawful upon some ground. Hereafter, presumption is to lighten the burden of the prosecutor. The victim will be spared the trouble of confessing and will go to his cell without mutilation or disquieting outcry.

He posed a question: "probably most of those accelerated to prison under the present Act will be unfortunate addicts and their abettors; but even they live under the Constitution. And where will the next step take us?" Finally, the Southern-born McReynolds observed that in 1914

> probably some drug containing opium could have been found in a million or more households within the Union. . . . Did every man and woman who possessed one of these instantly become a presumptive criminal and liable to imprisonment unless he could explain to the satisfaction of a jury when and where he got the stuff? Certainly, I cannot assent to any such notion, and it seems worthwhile to say so.[57]

The narrowness of the *Casey* decision, to say nothing of the bitterness involved, seems to have induced the Court to invite a "test case" that would resolve all the constitutional questions associated with the Harrison Narcotics Act. This case, *Nigro v. United States,* was decided in 1928. Chief Justice Taft wrote the majority opinion. Justices Stone, Sanford, Van Devanter, Holmes, and Brandeis joined Taft in this decision.[58] Once again, Justices Butler, Sutherland, and McReynolds dissented. The central question was simple: Was the Harrison Narcotics Act, when applied to persons not required to register, constitutional? Could the reach of the law go beyond the drug industry, pharmacists, and doctors? In its answer the Court again affirmed *Doremus,* asserting that the Harrison Narcotics Act was "obviously" a tax measure. To punish unregistered sellers was rationally related to tax collection, for it "keeps buying and selling on a plane where evasion of the tax will be difficult." Even punishing purchasers could be rationally related to tax collection, "because such drugs are not necessarily consumed by the purchaser but may be peddled or sold illegally" in avoidance of the tax.[59]

Thus, at the close of this contentious decade, the federal government had sustained most of its positions before the Court. The Harrison Narcotics Act had been sustained as a tax measure. Some limits on medical prescribing had been endorsed, with the *Linder* decision defining the boundaries of enforcement practice. Finally, prosecutions for possession had been made easier.

LAW ENFORCEMENT IN PRACTICE

The development of "official" policy in Washington, whether in the Treasury Department or the U.S. Supreme Court, has received most of the attention from historians. Whether the interest stems from a desire to map the changing dimensions of public policy or to examine national discourse on drugs, this approach tends to avoid close examination of the front lines of the emerging war on drugs. This section attempts to rectify this omission by characterizing the manner in which the federal government actually enforced its drug laws before 1930.

The Scope of Federal Law Enforcement

The fullest appreciation of the importance of federal drug law enforcement will only come when the extent of local drug enforcement is better documented. At present, there is relatively little published empirical data on local police activity against drug sellers and users. What is known strongly suggests that the initial enforcement of federal drug laws only supplemented local police efforts already well under way. Many states had state-level enforcement agencies, often pharmacy boards, and a handful of city police departments featured narcotic squads.[60]

New York City, home to more addicts than any other American city, developed its own system for dealing with drug users. Addicts received by the city Department of Correction were first sent to a clearinghouse hospital where they underwent a series of physical exams. Those suffering from contagious diseases stayed at the reception hospital, with those who had tuberculosis being sent to the tubercular hospital on Hart's Island. The majority of addicts, the "uncomplicated" cases, went directly to the huge jail complex at Rikers Island where they underwent a two-week gradual reduction process (for opiate addicts). For a week after completion of the withdrawal, addicts were allowed to do what they wished within the jail. After that time they advanced to light work or pulling weeds, then gradually began doing hard manual labor. After 100 drug-free days, the minimum under New York City laws, "the doctor is through, and the case becomes a sociologic problem, and the prevention of drug use [after release] is the paramount issue."[61] Later in the 1920s most addicts had been shifted to the Welfare Island penitentiary—over 1,100 addicts in 1930 alone.[62] Although no other U.S. city approached the scale of New York's efforts, the sum total of state and local efforts was almost certainly the equal of federal enforcement.

Although federal drug law enforcement was not the entire story, it did have special significance. First, while local agencies concentrated on arrests of addicts and low-level participants in the drug trade, federal enforcement raised the possibility of going after more prominent drug distributors. Though this potential was not often realized, the ability of federal agents to work independently across the country did result in a few notable cases being prosecuted. Second, the penalties for Harrison Narcotics Act violations were more severe than most state legislation; anyone convicted under the act was subject to as many as five years in prison. Most local drug prosecutions involved fines or short jail terms, and sentences to state prison terms seem to have been comparatively rare. Overall trends in federal drug law enforcement point to the influence of trends in official policy and the Supreme Court decisions. Table 2.1 indicates the number of convictions obtained under the Harrison Narcotics Act between 1915 and 1930.

The trends in convictions under the Harrison Narcotics Act appear to confirm that the first few years of the act were characterized by uncertain enforcement. Then 1919 marks the start of a vigorous upward

TABLE 2.1. Harrison Act Convictions by Year, 1915-1930

Year	Convictions
1915	106
1916	663
1917	445
1918	392
1919	582
1920	908
1921	1,583
1922	3,104
1923	4,194
1924	4,242
1925	5,600
1926	5,120
1927	4,469
1928	4,738
1929	5,193
1930	4,962

Sources: For 1915-1926, see *Annual Report of the Commissioner of Internal Revenue* (Washington, DC: GPO); for 1927-1930, see *Annual Report of the Commissioner of Prohibition* (Washington, DC: GPO). Each report covers the fiscal year ending June 30 of the reported year.

trend in the numbers of convictions, which continued until about 1925. Changes in the administration of Harrison Narcotics Act enforcement, coupled with the Supreme Court's blessing, appear to have allowed for far greater numbers of convictions. The leveling off of the number of convictions after 1925 seems to reflect the fact that few new enforcement resources were added after 1925.

The Targets of Federal Law Enforcement

Who were the targets of this enforcement activity? Judging from the amount of attention paid to the position of doctors under the law, one might well assume that federal prisons were bulging with physicians. Doctors were important targets, as we shall see, but the focus of federal drug law enforcement was elsewhere. One way to specify the nature of enforcement activity is to distinguish between the conviction of registered and unregistered persons, as seen in Table 2.2.

Long before the *Linder* decision, successful prosecutions of registered persons was rare, and it became no more common afterward. Prosecutions of unregistered persons were the rationale behind the overall trends. "Unregistered," of course, did not preclude physicians—unregistered doctors were prosecuted. Their numbers, however, were exceedingly small. In 1919 only twelve of 1,008 cases brought against unregistered persons were brought against physicians. In addition, only one drug manufacturer and fifty-five retail druggists were prosecuted, thus leaving 940 cases brought against persons entirely outside the medical/pharmaceutical fields.

Walter L. Treadway, head of the Mental Hygiene Division of the Public Health Service, reviewed enforcement activity for the third quarter of 1929.[63] Physicians constituted only a small proportion of arrests. Federal agents made 2,040 arrests during this period, of which 1,996 were unregistered persons. The forty-four registered arrestees included physicians, dentists, pharmacists, and veterinary surgeons. Some of the unregistered persons may have been doctors, but no more than a few; a review of the educational background of unregistered persons indicated that less than 1 percent had a professional degree of any kind.

Despite the promise of federal agents attacking the illicit trade at its core, most enforcement activity seems to have been at the margins of the drug marketplace.[64] One measure of activity focused on high-

TABLE 2.2. Harrison Act Cases by Type and Outcome, 1915-1930

	Convictions		Acquittals		Compromises	
Year	Reg.	Unreg.	Reg.	Unreg.	Reg.	Unreg.
1915	7	108	10	15	15	12
1916	83	580	20	163	555	29
1917	160	285	29	73	407	4
1918	86	396	42	41	274	23
1919	247	335	19	39	493	26
1920	134	774	40	89	515	31
1921	255	1,328	10	109	286	18
1922	159	2,945	33	199	498	17
1923	241	3,953	30	255	734	16
1924	245	3,997	24	252	735	23
1925	317	5,283	14	190	1,069	38
1926	223	4,835	12	189	1,965	54
1927	180	4,278	22	101	2,013	70
1928	148	4,577	3	105	1,182	38
1929	149	5,037	9	111	1,016	20
1930	202	4,760	9	164	1,098	16

Sources: For 1915-1926, see *Annual Report of the Commissioner of Internal Revenue* (Washington, DC: GPO); for 1927-1930, see *Annual Report of the Commissioner of Prohibition* (Washington, DC: GPO). Each report covers the fiscal year ending June 30 of the reported year. The government also reported pending cases and dropped cases, which are not reported here.

level drug traffic in the quantity of drugs seized. By almost any measure, these numbers were pitifully small. In most years federal agents seized between 6,000 and 26,000 ounces of illegal drugs, much of which would come from one or two large drug seizures. These numbers represent a tiny fraction of the quantity of drugs coming into the United States and suggest little or no meaningful enforcement activity against traffickers. Most large seizures were the result of chance discoveries by customs agents or anonymous tip-offs, not the penetration of the upper levels of drug distribution networks. As a consequence, drug seizures were seldom accompanied by arrests.[65]

Returning to 1929, the third-quarter data reveals similar patterns. Of the 1,996 unregistered persons arrested, fully 1,451 (72.7 percent) were themselves drug addicts. The crimes for which these addicts were charged suggests how marginal the subjects of federal enforcement were: possession (570 cases), sale (549), vagrancy related to drug addiction (95), drug addict (130), purchase (72), forging prescriptions (30), sending drugs through the mail (2), unlawful importation of drugs (2), transporting drugs (1).[66] Even when the non-addict arrests, most of which were for sales, are added to the total, just under half of all arrests were for possession, purchase, vagrancy, or merely for being an addict. This was the "hidden" drug war, waged on the most vulnerable groups within the illicit drug marketplace.[67]

COPS AND DOCTORS

If the federal government was not jailing, or even arresting, large numbers of doctors, how should the relationship between doctors and the government be characterized? Certainly, many physicians had joined forces with the government's war on drugs and gladly turned their backs on the addict and drug user. The third-quarter data for 1929 show only three arrests for "prescribing to addicts"—the very category that had been the subject of such heated debate over the previous decade.

The three arrests are somewhat misleading. If the *Linder* decision effectively ended most prosecutions of physicians for prescribing to addicts, federal agents were determined to ignore the spirit of the Court's ruling. As committed as ever to opposing any form of maintenance, the Treasury Department merely continued its ongoing practice of intimidation that fell just short of arrest. In the fiscal year ending in 1916, federal agents had formally pursued 21,844 cases against registered persons. Doctors constituted half of this number, dentists or veterinarians a quarter, and the remaining registered persons were druggists. Cases were dropped in 20,603 (94.3 percent) instances. Ten years later, 6,411 formal prosecutions were initiated against registered persons in the fiscal year ending in 1926. The largest number, 2,802 (43.7 percent), were "pending" at the end of the year, 1,409 (22 percent) were dropped, and 1,965 (30.7 percent) were settled "by compromise." Only 3 percent of these cases resulted in convictions.

Although few convictions were obtained, the "pending" and "compromise" categories suggest the ways federal agents could influence doctors. Physicians who were arrested for a violation of the Harrison Narcotics Act might, for example, simply be left with pending charges for an extended period of time. This would amount to a warning. Anything that could scare a doctor concerned drug agents. The use of the compromise was also a way of asserting federal authority without having to pursue a case in court. A compromise referred to a financial settlement reached between federal officials and a doctor before an indictment was sought. Not unlike a fine, the compromise was a device by which a doctor could avoid a formal prosecution. However, the compromise was also tantamount to a confession of guilt, and its widespread use provoked a furious reaction from the medical profession. A letter to the *Journal of the American Medical Association* denounced the compromise as "legalized blackmail." The AMA's legislative counsel, William C. Woodward, condemned "the distress and humiliation" faced by a physician who "must either compromise his case or go into a public court room and defend himself, with the possibility of spreading the suspicion that he is a 'dope peddler.'"[68]

The Treasury Department's list of "those under surveillance" in the third quarter of 1929 not only shows the limitation of relying solely on convictions to understand enforcement activity, but even the limitations of formal arrests: unregistered addicts (142), unregistered nonaddicts (17), registered addicts (61), registered nonaddicts (141).[69] The surveillance figures, with their emphasis on registered persons, strongly emphasize the fields of medicine and pharmacy, as do the "charges" that motivated the surveillance: failure to keep proper records, forging prescriptions, prescribing for addicts, improperly writing prescriptions, and filling unsigned prescriptions. Here, too, there are aspects of a "hidden" drug war—in this instance, a policy of harassment and intimidation without legal sanction.

PENETRATING THE MARKETPLACE

Over the course of this period, the problem of drug enforcement changed. At the outset, the illicit drug traffic was not well organized. Most drug sales involved a form of diversion from legal drug supply, with a small number of doctors and druggists supplying larger numbers of peddlers. Facing this type of distribution system, enforcement was comparatively easy. Doctors and druggists were readily identifiable, operated from fixed locations, and usually carried on a legitimate practice or business in addition to their illicit enterprises. Still easier targets were professionals such as Dr. Linder, who made no effort to conceal what they regarded as legitimate medical practice.

By the middle of the 1920s, the illegal market assumed its more modern form. Enforcement activities drove the first generation of doctor and druggist-sellers out of business, while federal legislation prohibiting heroin entirely, beginning in 1924, closed off one source of illegal drug diversion. Still more significant was the passage by Congress in 1922 of the Jones-Miller Act, which placed severe restrictions on the export of narcotics from the United States. To that point, one of the most significant sources of illegal drugs had been legally manufactured drugs exported to Canada or Mexico and then smuggled back into the United States. Replacing the old diversion system were far more sophisticated and well-organized drug trafficking networks.[70] By 1931 the National Commission on Law Observance and Enforcement concluded that the enforcement of prohibitory laws had created sophisticated systems of manufacture and distribution that posed enormous challenges for law enforcement.

The central challenge for law enforcement was to penetrate this effectively closed marketplace. Without a "victim," or at least a complainant, the burden of proof fell solely on drug agents themselves. In practice, this left agents with two options. The first was to confront sellers directly, a tactic that was almost inherently dangerous and violent. One of the few pre-Anslinger drug agents to write about his career, agent William J. Spillard, described many such confrontations. On a raid in Chicago's South Side, one woman pulled a revolver, held it to Spillard's head, and pulled the trigger twice, fortunately on the only two empty rounds in the chamber.[71] The key to understanding the confrontation is the importance of the physical evidence in drug cases—often agents and subjects were battling to secure the drugs themselves, without which a case could easily be lost.

The second option for agents was to secure inside information otherwise unavailable to them. In practice, this meant the use of informants from within the drug marketplace. Spillard wrote, "informers are a necessary evil in tracing down criminals. Commonly known as stool pigeons in the underworld, these men whom nobody is supposed to like are a narcotic agent's right arm in many instances."[72] Most informants were themselves addicts, their cooperation often secured through either the threat of arrest or the promise of a supply of narcotics. This latter offer proved especially helpful with addicts experiencing withdrawal while in custody. Alfred L. Tennyson, legal advisor to Harry Anslinger, head of the Federal Bureau of Narcotics, observed in 1930 that without the addict informant, "we could not make a case."[73]

CORRUPTION

From the outset, drug law enforcement had been riddled with corrupt practices. "Working under proto" was the phrase used by many dealers when they were able to induce police protection through payoffs. Even before the Harrison Narcotics Act, local police were heavily involved in drug enforcement-related corruption. Several factors accelerated these trends during the 1920s. First, although the close connections between drug agents and participants in the illicit market may have facilitated the penetration of the marketplace, they also facilitated corrupt relationships. Second, newer drug trafficking organizations had considerably more resources to devote to buying the allegiance, or at least the inactivity, of officials at all levels. Finally, the quasi-independent nature of drug agents, and drug squads at the local level, excused them from the traditional hierarchical structure of law enforcement and meant that most agents' activities were essentially unsupervised.[74]

Throughout the 1920s scandal followed scandal. In Chicago, agents were charged with accepting money from peddlers in order to turn them loose, and with selling drugs they had seized from dealers.[75] Corruption among his agents ended Levi Nutt's reign as head of the Narcotics Division in 1930. Arnold Rothstein, a well-known New York City gambler, had also masterminded the organization of one of the largest post-Harrison Narcotics Act drug trafficking organizations. His death in 1928 exposed his drug trafficking enterprise to public scrutiny, revealing the close ties between New York-based agents of the Narcotics Division and Rothstein's organization.[76]

DRUG AGENTS AND ROUTINE ACTIVITIES

Corruption doubtlessly interfered with the objectives of federal drug policy, but no more so than the passivity and routine that characterized most enforcement practices. By the end of the 1920s, drug enforcement in the United States had settled into a routine in which agents worked the same informants continually to produce enough low-level arrests to satisfy their superiors. The nature of the addict population made this task even easier, for the "epidemic" of new heroin use was long over, leaving most cities with a relatively stable corps of long-term addicts. Half of the addicts who passed through New York's Welfare Island Penitentiary in 1930 had been addicted for at least a decade. Indeed, Welfare Island in 1930 housed a number of addicts whose use predated the Harrison Narcotics Act itself, at least 10 percent of the total. At the Narcotic Ward at Bellevue Hospital in 1928 and 1929, the average age was thirty-four, with an average length of addiction being eleven years.[77] A simple comparison shows the trend. Table 2.3 indicates the ages of addicts who reported to the New York City Narcotic Clinic in 1919 and 1920, and the ages of addicts arrested for violations of the Harrison Narcotics Act in 1929.

TABLE 2.3. Drug Addicts by Age

	NYC, 1919-1920		Federal Arrests, 1929	
Age	**Number**	**Percent**	**Number**	**Percent**
15-19	743	10	12	1
20-24	2,142	29	139	8
25-29	2,218	30	281	17
30-34	1,155	16	322	19
35-39	766	10	297	18
40-50	365	5	312	19
50+	75	1	191	12

Sources: C. Edouard Sandoz, "Report on Morphinism to the Municipal Court of Boston," *Journal of Criminal Law and Criminology* (1922), 10-55. W. L. Treadway, "Further Observations on the Epidemiology of Narcotic Drug Addiction," *Public Health Reports* 45 (March 14, 1930), 541-553.

While the arrest data are national, with only 10 percent being from New York, and reflect some Narcotics Division efforts to root out older morphine addicts in other parts of the country, the differences are striking. Other studies confirm the image of an aging and stable addict population in the 1930s, readily subject to drug agent surveillance.[78] Federal drug enforcement settled into a comfortable routine that even the bureaucratic reorganization of 1930 would not entirely disturb. This routine largely meant arrests of drug retailers and their customers, with relatively little attention paid to the sources of supply or the conditions that allowed the trade to thrive in the first place.[79]

URLs FOR PRIMARY DOCUMENTS

Interstate Commerce Act of 1887
<http://www.civics-online.org/library/formatted/texts/interstate_commerce.html>

Sherman Anti-Trust Act of 1890
<http://voteview.uh.edu/antitrst.htm>

Mann Act, 1910
<http://www.fau.edu/~tunick/courses/conlaw/mann.html>

Espionage Act, June 15, 1917
<http://www.staff.uiuc.edu/~rcunning/espact.htm>

U.S. Sedition Act, May 16, 1918
<http://www.lib.byu.edu/~rdh/wwi/1918/usspy.html>

"More than 1,000,000 Drug Users in U.S.," June 13, 1919, *The New York Times*
<http://www.druglibrary.org/schaffer/history/e1910/nyt061310.htm>

Finding Aid to the Lawrence Kolb Papers, 1912-1972
<http://www.nlm.nih.gov/hmd/manuscripts/ead/kolb.html>

"Two Federal Prison Farms to Care for Drug Addicts," April 14, 1929, *The New York Times*

<http://www.druglibrary.org/schaffer/history/e1920/federalprisonfarms.htm>

United States v. Coca Cola Company of Atlanta, May 22,1916 <http://www.druglibrary.org/schaffer/legal/11910/ united_states_v_coca_cola.htm>

McGinis et al. v. People of the State of California, May 20, 1918 <http://www.druglibrary.org/schaffer/legal/11910/mcginis_v_california.htm>

Hammer v. Dagenhart, June 3, 1918 <http://www.agh-attorneys.com/4_hammer_v_dagenhart.htm>

Webb et al. v. United States, March 3, 1919 <http://www.druglibrary.org/schaffer/history/webb.html>

U.S. v. Doremus, March 3, 1919 <http://www.druglibrary.org/schaffer/history/doremus.html>

State of Minnesota on the Relation of *Whipple v. Martinson,* Sheriff of Hennepin County, Minnesota <http://www.druglibrary.org/schaffer/legal/11920/Whipple.htm>

United States v. Balint et al., March 27, 1922 <http://www.druglibrary.org/schaffer/legal/11920/ united_states_v_balint_et_al.htm>

United States v. Wong Sing <http://www.druglibrary.org/schaffer/legal/11920/Wongsing.htm>

United States v. Behrman, March 27, 1922 <http://www.druglibrary.org/schaffer/legal/11920/united_states_v_behrman.htm>

Child Labor Tax Act, 1922 <http://www.eco.freedom.org/ac92pg1512.shtml>

Linder v. United States, April 13, 1925 <http://www.druglibrary.org/schaffer/history/linderv.htm>

Yee Hem v. United States, April 27, 1925
<http://www.druglibrary.org/schaffer/legal/11920/Yeehem.htm>

Agnello et al. v. United States, October 12, 1925
<http://www.druglibrary/schaffer/legal/11920/
agnello_vs_us_us_supreme_court.htm>

United States v. Daugherty, January 4, 1926
<http://www.druglibrary.org/schaffer/legal/11920/Daugherty.htm>

Boyd v. United States, April 19, 1926
<http://www.druglibrary.org/schaffer/legal/11920/Boyd.htm>

Wong Tai v. United States, January 3, 1927
<http://www.druglibrary.org/schaffer/legal/11920/Wongtai.htm>

Alston v. United States, May 16, 1927
<http://www.druglibrary.org/schaffer/legal/11920/ altson_v_us_
us_supreme_court.htm>

Casey v. United States, April 8, 1928
<http://www.druglibrary.org/schaffer/legal/11920/ casey_v_us_
us_supreme_court_1928.htm>

Nigro v. United States, April 9, 1928
<http://www.druglibrary.org/schaffer/legal/11920/Nigrovus.htm>

NOTES

1. *United States v. Jin Fuey Moy,* 241 U.S. 394 (1916).

2. For very useful accounts of World War I and its domestic impact, see Robert H. Zieger, *America's Great War: World War I and the American Experience* (New York: Rowman and Littlefield, 2000) and David M. Kennedy, *Over Here: The First World War and American Society* (New York: Oxford University Press, 1980).

3. David F. Musto, *The American Disease: Origins of Narcotic Control,* Revised Edition (New York: Oxford University Press, 1987), pp. 143-144. Roper had been first assistant postmaster general in Wilson's first term. After aiding Wilson's re-election bid, Roper received the appointment as head of the Bureau of Internal Revenue. In the next Democratic administration (Franklin D. Roosevelt's) Roper served as secretary of commerce.

4. Musto, *The American Disease,* p. 145. No federal commitment to providing treatment to addicts was forthcoming until the Porter Bill passed in 1929. The legislation authorized the construction of two narcotics hospitals that were opened in 1935 (Lexington, Kentucky) and 1938 (Fort Worth, Texas), respectively. Even then, these institutions were constructed primarily with the objective of relieving the federal prisons of their addict prisoners.

5. Ernest Bishop, in *The Narcotic Drug Problem* (New York: Macmillan, 1920), p. 58, wrote that addiction "changed the body's physiology in such a way that opiates were necessary for the patient to remain normal." See also Edward Huntington Williams, *Opiate Addiction: Its Handling and Treatment* (New York: Macmillan, 1922).

6. V. V. Anderson, "Drug Users in Court," *Journal of Criminal Law and Criminology* (March 1917): 903-906.

7. Lawrence Kolb, "Pleasure and Deterioration from Narcotic Addiction," *Mental Hygiene* 9 (1925): 699-724.

8. H. S. Cumming, "Control of Addiction Mainly a Police Problem," *American City Magazine* (November 1925): 509.

9. Lawrence Kolb Sr. and A. G. DuMez, *The Prevalence and Trend of Drug Addiction in the United States and Factors Influencing It* (Washington, DC: GPO, 1924), p. 23. Years later, Alfred Lindesmith recalled the "remarkable agreement between the medical officials of the Public Health Service and the police officials of the Federal Bureau of Narcotics." Lindesmith, *The Addict and the Law* (Bloomington, IN: Indiana University Press, 1965).

10. Alfred C. Prentice, "The Problem of the Narcotic Addict," *JAMA* 76 (1921): 1553. The American Public Health Association's Habit Forming Drugs Committee, on the other hand, fully supported the clinics. On the committee were such leading clinic managers and maintenance advocates as Charles Terry, Ernest Bishop, Oscar Dowling, and Lucius Brown.

11. Thomas S. Blair, "The Dope Doctor: And Other Country Cousins of the Moonshiner," *The Survey* (April 3, 1920): 20. The letter from the Pennsylvania critic was reprinted within *The Survey* article.

12. Ibid.

13. The Narcotic Division of the Prohibition Unit remained in that bureaucratic arrangement until 1930, when the Narcotic Division was separated and placed under the leadership of Harry J. Anslinger. See Chapters 3 and 4 in this book.

14. Musto, *The American Disease,* pp. 146-149.

15. "A Municipal Drug Addict Clinic," *The American City* 18 (May 1918): 437.

16. Paul W. Kearney, "Common Sense in Drug Control," *The American City* 24 (1921): 499-503.

17. For more on these principles in action in the Anslinger era, see Chapters 3 and 4 in this book. Also, Jim Baumohl, "Maintaining Orthodoxy: The Depression-Era Struggle over Morphine Maintenance in California," *Contemporary Drug Problems* 27 (spring 2000).

18. *Special Committee of Investigation, Appointed March 25, 1918, by the Secretary of the Treasury: Traffic in Narcotic Drugs* (Washington, DC: GPO, 1919).

19. Musto, *The American Disease,* p. 330.

20. There are too many articles using the 1 million figure to review here. Representative articles include "Drug Addicts in America," *The Outlook* (June 25, 1919):

315; "Drug Prohibition," *The Survey* (February 22, 1919): 727-728; "The Drug Menace in America," *American Review of Reviews* (September 1919): 331. The last of these offered an independent estimate of 1.5 to 5 million addicts in the United States and observed, "In New York City estimates of drug habitues have doubled within a period of months to 200,000."

21. "A Million Drug Fiends," *The New York Times,* September 13, 1918.

22. Thomas S. Blair, "The Dope Doctor," 19-20.

23. Pearce Bailey, "The Drug Habit in the United States," *The New Republic,* March 16, 1921, pp. 67-69.

24. The most thorough examination of this question may be found in David Courtwright, *Dark Paradise* (Cambridge, MA: Harvard University Press, 1982).

25. Lawrence Kolb Sr. and A. G. DuMez, "The Prevalence and Trend of Drug Addiction in the United States and Factors Influencing It," *Public Health Reports* 39 (May 1924): 1179-1204.

26. C. L. Eddy, "One Million Drug Addicts in the United States," *Literary Digest* 78 (August 25, 1923): 22-23.

27. The best source for information concerning Hobson's mid-1920s activities, and government reaction, are the Lawrence Kolb Papers, 1912-1972, located in Modern Manuscripts Collection, History of Medicine Division, National Library of Medicine Bethesda, MD, MS C 279 (hereafter Kolb Papers).

28. The bill was S 5204, submitted in the second session of the 69th Congress. The Public Health Service worked against this bill. See H. S. Cumming, surgeon general to Senator James Couzens, January 25, 1927, Kolb Papers. Senator Couzens headed the Senate Committee on Education and Labor, to which the bill was directed.

29. W. A. Swanberg, *Citizen Hearst* (New York: Scribner, 1961), pp. 556-557. Charles Beard, speaking in 1935, said that "William Randolph Hearst had pandered to depraved tastes and has been an enemy of everything that is noblest and best in our American tradition. . . . There is not a cesspool of vice and crime which Hearst has not raked and exploited for money-making purposes." It should also be noted that Senator Royal S. Copeland was a good friend of Hearst's. In the post-Hobson era, Hearst was a more reliable ally, especially to Harry J. Anslinger and the Federal Bureau of Narcotics. Hearst was instrumental in pushing for the reorganization of 1930 that ended the Narcotic Division and replaced Nutt with Anslinger. In the late 1930s, Hearst's Seattle *Post-Intelligencer* helped the FBN to defeat a narcotic clinic proposal in the state.

30. Lawrence Kolb to Winifred Putnam (April 1, 1926), Kolb Papers.

31. See H. S. Cumming to Senator James Couzens (January 25, 1927), Kolb Papers; Levi Nutt, "Memorandum for the Secretary of the Treasury Concerning Richmond P. Hobson and Drug Addiction," Kolb Papers.

32. As late as 1930, however, the "1 million addicts" figure still had not been routed. That year U.S. Representative Stephen Porter, one of the leading figures in drug policy circles, testified before a congressional committee that there was "no possible way" to accurately determine the number of drug addicts in the country. Porter noted, however, "some authorities say a million." House Committee on Ways and Means, *Bureau of Narcotics,* 71st Congress, 2nd Session, 1930, p. 15.

33. Alfred R. Lindesmith, *The Addict and the Law* (Bloomington, IN: Indiana University Press, 1965), p. 257. While Hobson was never personally destroyed by

the federal government, the Federal Bureau of Narcotics subjected antimarijuana crusader Earl Albert Rowell to a brutal series of attacks.

34. Among the lesser-known congressional actions was a 1912 ban on the interstate transportation of prizefight films.

35. 267 U.S. 432 (1925). The Mann Act was upheld in *Hoke v. United States,* 227 U.S. 308 (1913).

36. 247 U.S. 251 (1918); Bernard Schwartz, *A History of the Supreme Court* (New York: Oxford University Press, 1993), pp. 212-213.

37. The case is formally known as *Bailey v. Drexel Furniture Company,* 257 U.S. 20 (1922); Schwartz, *A History of the Supreme Court,* pp. 217-218.

38. Musto, *The American Disease,* pp. 128-130.

39. *U.S. v. Doremus,* 249 U.S. 86 (1919). President Theodore Roosevelt appointed William Rufus Day to the Court in 1903 following four years as a U.S. Circuit Court of Appeals judge. Previously, Day had been an assistant secretary of state under President McKinley.

40. *U.S. v. Doremus,* 249 U.S. 86 (1919) at 94.

41. See, for example, *Alston v. United States,* 274 U.S. 289 (1927). It appears that at least some in government sought other legal foundations for drug law enforcement. One agent seized opium bowls and pipes under federal obscenity laws, arguing that "opium smoking stimulates sexual desire" and that smoking paraphernalia were therefore obscene. Levi Nutt wrote to Kolb, asking him what he thought of the idea. Kolb's reply was not encouraging. Lawrence Kolb to Levi Nutt (March 28, 1924), Kolb Papers.

42. *Casey v. United States,* 276 U.S. 413 (1928).

43. *Nigro v. United States,* 276 U.S. 332 (1928). Data on the revenues generated by the Harrison Act taxes present a mixed picture. Tax revenues certainly grew after 1918, from less than $200,000 to over $1,000,000 in 1920. Tax revenues were declining by 1928, however, and total receipts were actually lower in 1928 than in 1919, the year *Doremus* was decided.

44. *Webb et al. v. United States,* 249 U.S. 96 (1919).

45. *U.S. v. Doremus,* 249 U.S. 86 (1919).

46. Schwartz, *A History of the Supreme Court,* p. 210.

47. The Court, and subsequent discussions of the case, made much of the fact that Behrman had prescribed "three thousand" doses of morphine. In fact, he had done no such thing. The Court based this number, presumably taken from the government's own presentation, on the standard dose prescribed in *Wood's United States Dispensatory.* That dose, .25 grains, was far below what most addicts used. Reports from the narcotic clinics indicated that they dispensed an average of 7.25 grains to addicts, while initial doses at some clinics went as high as 15 grains. Using these doses, Behrman had actually dispensed between fifty and 100 daily doses—a great deal, to be sure, but not so much as to totally defeat his argument that this was intended for patient self-administration over an extended period of time.

48. *U.S. v. Behrman,* 258 U.S. 280 (1922).

49. *Linder v. U.S.,* 268 U.S. 5 (1925), at 16.

51. Ibid., at 18.

52. Lindesmith, *The Addict and the Law,* p. 11.

53. *United States v. Wong Sing,* 260 U.S. 18 (1922); *Yee Hem v. United States,* 268 U.S. 178 (1925).

54. *Casey v. United States,* 276 U.S. 413 (1928).

55. Ibid., at 418.

56. Ibid., at 426-429.

57. Ibid., at 420-421. The other dissenters were Brandeis and Sanford. Brandeis in particular disavowed the Butler dissent, instead focusing on what he felt was a clear instance of entrapment and police misconduct. As Brandeis wrote: "I am aware that courts—mistaking relative social values and forgetting that a desirable end cannot justify foul means—have, in their zeal to punish, sanctioned the use of evidence obtained through criminal violation of property and personal rights or by other practices of detectives even more revolting. But the objection here is of a different nature . . . the alleged crime was instigated by officers of the Government . . . [the Government] may not provoke or create a crime and then punish the criminal, its creature." *Casey v. United States,* 276 U.S. 413 (1928), at 421-424.

58. *Nigro v. United States,* 276 U.S. 332 (1928). For a further look at the Court's criticisms of the Harrison Act leading up to the *Nigro* case, see "United States Supreme Court Distrusts Harrison Narcotic Act," editorial, *JAMA* 86 (1926): 627-628.

59. *Nigro v. United States,* 276 U.S. 332 (1928), at 344-346.

60. See Chapter 1 in this book.

61. James A. Hamilton, "The Treatment of Drug Addiction at the Correction Hospitals in New York City," *Journal of Criminal Law and Criminology* (May 1922): 122-126.

62. Joseph Fulling Fishman and Vee Terrys Perlman, "The Real Narcotic Addict," *American Mercury* 25 (January 1932): 100-107.

63. Walter L. Treadway, "Further Observations on the Epidemiology of Narcotic Drug Addiction," *Public Health Reports* 45 (March 14, 1930): 541-553.

64. The failure to deal with drug smuggling, for example, occurred despite passage of the Narcotic Drugs Import and Export Act in 1922. More often referred to as the Jones-Miller Act (after its House and Senate sponsors), the 1922 law placed much closer restrictions on the traffic of opiates and cocaine in and out of the country. It also introduced penalties of up to ten years in prison.

65. Charles Tuttle, U.S. attorney for the southern district of New York, testified to Congress in 1930 that no recent seizures had been accompanied by arrests. House Committee on Ways and Means, *Bureau of Narcotics,* 71st Congress, Second Session, 1930, p. 64.

66. Treadway, "Further Observations."

67. Indicative of the focus of federal law enforcement, the idea of attacking high-level trafficking networks seems not to have occurred to the Narcotics Division until shortly before its dissolution in 1930. The Federal Bureau made some limited progress against smugglers in 1930, but efforts remained scattershot and limited even then.

68. House Committee, *Bureau of Narcotics,* pp. 78-80. The letter is cited by Dr. Woodward and reprinted in full on page 79 of the committee hearings, though no publication information is given.

69. Treadway, "Further Observations."

70. Kathryn Meyer and Terry Parssinen, *Webs of Smoke: Smugglers, Warlords, Spies, and the History of the International Drug Trade* (New York: Rowman and Littlefield, 1998).

71. William J. Spillard, *Needle in a Haystack: The Exciting Adventures of a Federal Narcotic Agent* (New York: McGraw-Hill).

72. Ibid., p. 141.

73. House Committee, *Bureau of Narcotics,* p. 95.

74. The National Commission on Law Observance and Enforcement, *Enforcement of the Prohibition Laws,* Volume 4 (Washington, DC: GPO, 1931).

75. Joseph Spillane, "The Making of an Underground Market: Drug Selling in Chicago, 1900-1940," *Journal of Social History* 32 (fall 1998): 27-47.

76. See Jill Jonnes, *Hep-Cats, Narcs, and Pipe Dreams: A History of America's Romance with Illegal Drugs* (New York: Scribner, 1996), Chapter 5.

77. "Report of the Mayor's Committee of Correction, New York City," *American Journal of Psychiatry* 10 (November 1930): 433-538.

78. One might use, for example, data collected in 1936 and published in Michael J. Pescor, "A Statistical Analysis of the Clinical Records of Hospitalized Drug Addicts," *Public Health Reports,* Supplement 143 (Washington, DC: GPO, 1938). The stability of the addict population in the 1930s would end after World War II, with the start of successive waves of new heroin use. Compare the data collected by Pescor with, for example, W. G. Smith, E. H. Ellinwood Jr., and G. E. Vaillant, "Narcotic Addicts in the Mid-1960s," *Public Health Reports* 81 (May 1966): 403-412.

79. These conclusions echo those from my study of local antidrug efforts in Chicago for the same period. See Spillane, "The Making of an Underground Market."

Chapter 3

Under the Influence: Harry Anslinger's Role in Shaping America's Drug Policy

Rebecca Carroll

In the formation of U.S. drug policy, the period from 1930 to 1960 was unique because a single individual dominated it. Harry J. Anslinger was commissioner of the Federal Bureau of Narcotics (FBN, now Drug Enforcement Administration) from the bureau's inception in 1930 until his retirement in 1962. Probably more than any other person, Anslinger influenced Americans' attitudes toward narcotic drugs and drug users and sellers, depicting both users and sellers as criminals. Thus, from the 1930s, the United States emphasized criminal justice rather than public health in its approach to the problem of illicit narcotics. For his entire tenure, Anslinger responded to the nation's drug problem by arguing for long prison sentences, high fines, and compulsory hospitalization. Although some dissenters spoke out through the years, Anslinger remained America's predominant voice in the area of narcotics (see Photo 3.1).

Harry Jacob Anslinger was born on May 20, 1892, in Hollidaysburg, a small town in Pennsylvania. His parents, Robert John and Christina Fladtt Anslinger, emigrated from Switzerland, eventually settling in Hollidaysburg, where Robert opened a barbershop and later worked for the Pennsylvania Railroad.[1] One explanation for Anslinger's attitude toward narcotic drugs was that he was born during the Progressive Era in the United States, the period from approximately the 1890s through the first two decades of the twentieth century. Thus, this period influenced the formative decades of Anslinger's life. For the first- and second-generation rural American, the

PHOTO 3.1. Young Harry J. Anslinger poses for a formal portrait. (*Source:* Harry J. Anslinger Collection, courtesy of Historical Collections and Labor Archives, Special Collections Library, The Pennsylvania State University [Folder #3]; photo restored by Barry Eckhouse)

city was the locus of the corruption that the Progressives sought to expose and eliminate. Historian Richard Hofstadter wrote,

> The whole cast of American thinking in this period was deeply affected by the experience of the rural mind confronted with the phenomena of urban life, its crowding, poverty, crime, corruption, impersonality, and ethnic chaos. To the rural migrant, raised in respectable quietude and the high-toned imperatives of evangelical Protestantism, the city seemed not merely a new social form or way of life but a strange threat to civilization itself.[2]

An event early in Anslinger's life may help to explain further his attitude toward narcotic drugs. In his book, *The Murderers,* Anslinger wrote,

> As a youngster of twelve, visiting in the house of a neighboring farmer, I heard the screaming of a woman on the second floor. I had never heard such cries of pain before. The woman, I learned later, was addicted, like many other women of that period, to morphine, a drug whose dangers most medical authorities did not yet recognize. All I remember was that I heard a woman in pain, whose cries seemed to fill my whole twelve-year-old being. Then her husband came running down the stairs, telling me I had to get into the cart and drive to town. I was to pick up a package at the drug store and bring it back for the woman.
>
> I recall driving those horses, lashing at them, convinced that the woman would die if I did not get back in time. When I returned with the package—it was morphine—the man hurried upstairs to give the woman the dosage. In a little while her screams stopped and a hush came over the house.
>
> I never forgot those screams. Nor did I forget that the morphine she had required was sold to a twelve-year-old boy, no questions asked.[3]

Anslinger's lifetime work with drug traffic and drug addiction was forever linked to Progressivism's tension between the old and the new, the early immigrant and the new immigrant, the small town and the city, and the minority and the majority. When almost seventy years old, Anslinger described his hometown, revealing his ingrained impression of the fracture between the old and the new:

> As I grew up, I saw other glimpses of drug addiction and its effects in this community, this small-town symbol of Main Street America of that era, with its mixture of old families and new immigrants, rolling farmlands and new factories, miners and roadworkers, foremen and factory heads.[4]

From September 16, 1913, to August 3, 1915, Anslinger attended Pennsylvania State College, now The Pennsylvania State University, receiving a certificate of completion for a two-year program in the School of Agriculture.[5] Though Anslinger's career prior to becoming commissioner was inconspicuous, it revealed some inclination for criminal investigations and undercover work. His early work experience showed rapid promotion, and it provided evidence that, as a young man, Anslinger was entrusted with duties requiring increasing sensitivity and diplomacy. From 1916 to 1921 he held three positions. The first was in the Fire Marshal's Office of the Pennsylvania State Police Department in Harrisburg, the second as a clerk with the War Department, Inspection Division of the Ordnance Department in Washington, DC, and the third as a clerk with the State Department at the American Legation, The Hague, Netherlands. On October 26, 1921, Anslinger began a six-year assignment with the U.S. State Department as a vice consul. His first assignment as American consul took him to Hamburg, Germany, where a former State Department co-worker said that Anslinger "first became deeply interested in narcotic control, working on cases in which United States seamen jumped ship overseas and became involved in various criminal offenses, including narcotics."[6] About his work in Hamburg, Anslinger wrote,

> In this post I helped repatriate many American seamen. Young Americans, most of them, young fellows whose faces bore the stamp of the opium smoker, the user of morphine or the new "kick" called heroin. I saw it also on the skeleton faces of men of other countries, seeking visas or other help from us.
>
> You always knew—and always with a stab of pity. The loathing that had begun in me as a boy in Pennsylvania increased as this blight was revealed in so many faces.[7]

In 1929, Anslinger became assistant commissioner of Prohibition, which he called "a thankless and impossible assignment."[8] Yet he wrote, "So long as I held this post, I labored to my fullest strength to enforce and make effective the anti-alcohol laws of the United States."[9] The Prohibition Bureau worked diligently to enforce antialcohol laws, but the majority of Americans rejected Prohibition. According to Anslinger, the American people had an appetite for alcohol, and "criminal gangs were feeding this appetite. Liquor poured across the borders not in a trickle but in a flood."[10]

However, as Prohibition in the United States ended, criminal gangs found another market in which to amass their fortune: narcotic drugs. Traffic in this vice provided Anslinger with employment for the next thirty-two years. On June 9, 1930, President Herbert Hoover signed HR 11143, creating a Narcotics Bureau within the Treasury Department with a commissioner to be named by the president and approved by Congress.[11] At the bureau's inception on July 1, 1930, Secretary of the Treasury Andrew Mellon appointed Anslinger acting commissioner.[12] While he was a skillful but low-level bureaucrat in the early years of his government career, neither his academic training nor his previous work experience sufficiently prepared him to be the commissioner of the FBN, a position that involved dealing with medical and legal issues, as well as directing a minimum of 200 employees (see Photo 3.2).

Anslinger claimed not to have sought actively the permanent position, a notion substantiated by one of his senior agents, Colonel Garland Williams, who said, "He did not want that job. He wanted to be a diplomat." According to Williams, because marijuana was entering the United States from Central and South America, the Treasury Department was faced with a "new law enforcement activity that was beyond the usual scope of police policies and practices." He said, "Leadership and training required close cooperation with drug authorities in many foreign countries. The Treasury Department needed an internationalist to run the Narcotics Bureau. Anslinger was already neck-deep in the Coast Guard and Customs antismuggling activities involving liquor, heroin from the Middle East through Italy, France, and Florida, and *Cannabis sativa* from Central and South America."[13]

For the next three decades, through the depression, World War II, McCarthyism, and five presidential administrations, Harry Anslinger led the charge against illicit narcotic drugs in the United States. Many legislators, law enforcement officials, the press, and the public regarded him as the nation's expert on drugs. In fact, numerous entries in the *Congressional Record,* contemporary journals, and letters in Anslinger's files contained references to him as "the world expert" on narcotic drugs.

From his experience in the Prohibition Bureau, Anslinger knew that controlling narcotics would be difficult. From the very beginning, his bureau was challenged, even from inside the government. In *The Murderers,* Anslinger recalled,

> I had been in office only a few weeks when I found myself and the new Bureau the object of a blistering assault from the floor of the Senate.
>
> Senator Cole Blease of South Carolina had purchased opium

> through an undercover private detective he had hired, to prove how easy it was to buy dope in the capital of the United States. The Senator now rose from his seat, waving in his hand the tin of opium. "This was purchased," he shouted, "only one block from where we are now deliberating."
>
> I knew that what he said was true. I had been working on the case since my first day in office. With this blunt charge from the floor of the Senate, however, time was running out—if the newly formed Federal Narcotics Bureau was to win and hold the respect and support of Congress and the public it would have to act fast.[14]

Although Anslinger had no medical or scientific background, and his bureau performed very few narcotics studies, he was a powerful and persuasive speaker, and he and his agents fabricated horror stories connecting drug use with violent crime. They presented these unsubstantiated stories as testimony before numerous congressional committees. Such evidence weighed heavily in the passage of at least three major pieces of drug legislation: the Marihuana Tax Act of 1937, the Boggs Act of 1951, and the Narcotic Control Act of 1956. Following are accounts of how the first two bills became law; the third is discussed in Chapter 4.

THE MARIHUANA TAX ACT OF 1937 (MTA)

The Harrison Narcotics Act of 1914, enforced by the Bureau of Internal Revenue in the Treasury Department, provided for federal control of narcotic drugs, but did not include marijuana because of its few positive medical and industrial uses: pharmacists used it as an analgesic in corn plasters and as an ingredient in mild sedatives; veterinarians used it until they were given access to more potent drugs; the bird seed industry manufactured it as food for canaries; and the Sherwin-Williams paint company used it as a drying agent.[15] Some scholars believe that marijuana was not included because the pharmaceutical industry opposed restricting so mild a substance.[16] Therefore, upon establishment of the FBN within the Treasury Depart-

PHOTO 3.2. Harry J. Anslinger stands with confiscated drugs. (*Source:* Harry J. Anslinger Collection, courtesy of Historical Collections and Labor Archives, Special Collections Library, The Pennsylvania State University [Folder #3])

ment, the bureau was not required to become involved in policing marijuana.

This was a blessing for Anslinger and the FBN because, for them, marijuana was a nuisance drug. Anslinger, a voracious reader, particularly on topics dealing with narcotic drugs, was probably aware of the National Wholesale Druggists' Association's (NWDA) belief that cannabis should not be included in any federal antinarcotic law because it "was not what might be called a habit-forming drug." NWDA chairperson Charles A. West gave that testimony at a House Ways and Means Committee hearing in 1911, and his testimony was substantiated by Albert Plaut of the pharmaceutical firm of Lehn and Fink, who said that opium and cocaine addicts would simply not find cannabis attractive because its effects were so different. In fact, before World War I, no one, not even the reformers, targeted cannabis as a problem in the United States.[17]

Further, having worked for the defunct Prohibition Bureau, Anslinger knew that judges were reluctant to fill their court calendars with what the courts considered minor offenses. Just as possession-of-alcohol cases were once considered nuisances, possession-of-marijuana cases, particularly in federal courts, would simply annoy the judges. Therefore, Anslinger concentrated the bureau's efforts on controlling the then most dangerous drugs—cocaine and opiates.[18]

Thus, for Anslinger and the FBN, federal enforcement of a marijuana law would have meant spreading a small staff of agents even more thinly. It would have exhausted FBN resources that were already reduced because of the depression. Finally, it would have taken agents and funds away from much more serious drug-related problems in and out of the United States.

During the first half of the 1930s, Anslinger gave very little attention to the issue of marijuana, speaking about it infrequently and usually to encourage individual states to pass the Uniform State Narcotic Drug Act, relegating marijuana jurisdiction to the state level. Anslinger claimed that the Uniform Act

> will close the gap between Federal and State laws, through which so many criminals escape well-merited punishment; it will eliminate suffering and pain; . . . it will prevent those States in which it is enacted from becoming a haven for criminal narcotic addicts and a dumping ground of narcotic peddlers.[19]

Anslinger's public statements during those years reflected his opinion that marijuana was indeed a problem on the state level, yet not enough of a problem to involve his federal agency. This was a delicate task: if he stressed the problem side too heavily, the public would demand federal protection against marijuana.

On December 14, 1935, during the House Appropriations hearings for fiscal year 1937, marijuana received little attention. Anslinger mentioned it briefly when asked about the Uniform Law, saying, "That has been of tremendous help, particularly with respect to the drug that we have no control over, marijuana. The traffic in that drug is increasing greatly. During the past harvesting season, in cooperation with agents of several States and local police officers, we destroyed more marijuana acreage than ever before," implying that the FBN was, perhaps generously, assisting states in eradication efforts wherever it could.[20]

When Representative Louis Ludlow of Indiana asked, "What drugs are comprehended when we use the general term, narcotics?" Anslinger responded, "Morphine, heroin, cocaine, and opium." When asked if marijuana caused drug addiction, Anslinger replied, "I do not think so."[21] Later Ludlow asked, "Heroin is the most dangerous of all drugs?" and Anslinger replied, "Absolutely."[22]

Until the end of 1935, Anslinger mostly ignored the topic of marijuana. However, if he had known what was to happen only a month into 1936, he might have offered quite different information.

On January 30, 1936, F. W. Russe, secretary of Mallinckrodt Chemical Works, wrote the following letter to Anslinger:

> I have been advised that House Bill 10586 introduced a few days ago, has a hidden rider or clause that will do away with the Federal Narcotic Department. I have not yet seen the bill but have written for a copy. If my information is correct can you advise me what is behind the bill?[23]

Known as the Secret Service Reorganization Act, the bill was introduced by Representative Robert L. Doughton, chair of the House Committee on Ways and Means, at the request of Secretary of the Treasury Henry Morgenthau Jr. and with the approval of President Roosevelt. Although the bill seemingly intended for the Treasury Department simply to run more efficiently, a provision in the bill would create a Secret Service Division whose chief, appointed by the Secretary of the Treasury, would assume responsibility for the Enforcement Division of the Alcohol Tax Unit, the Intelligence Unit of the Internal Revenue Service, the Customs Agency Service, and the Bureau of Narcotics.[24] This provision would presumably force the heads of those agencies, including Anslinger, out of their positions.

Immediately after Doughton introduced the Reorganization Act in January 1936, Anslinger's public statements changed. He began to discuss marijuana publicly. He seemed almost to forget about his previous delicate task of instilling just enough fear in his listeners to effect passage of the Uniform State Law but to stop short of federal legislation. Instead, Anslinger declared war on marijuana.

The following month, in a February 24, 1936, NBC radio address titled "The Need for Narcotic Education," Anslinger gave a brief history of narcotic legislation in regard to opium and heroin as his entree into discussing the need for the Uniform State Narcotic Drug Act. The Uniform Act, he said, was "designed to coordinate enforcement machinery through mandatory cooperation of state and federal officers, and to close the gap between Federal and State laws through which many criminals have escaped or evaded punishment." The last two pages of his seven-page address outlined the advantages of the Uniform Law exclusively concerning the marijuana "menace." In advocating the Uniform Law, Anslinger referred to marijuana: "The habit appears to be gaining ground rapidly. The gravity of this menace, the pernicious effects of the use of marijuana, and the necessity for its eradication, should be brought forcibly before the notice of our people."[25] This is a much different characterization of the situation than in his speech fourteen months earlier in December 1934, at the attorney general's Conference on Crime, in which he claimed that individual states were, in fact, "stamping out the use of the drug."[26] In the more recent speech, Anslinger characterized marijuana use as a "situation . . . full of danger." Although the situation could have changed over fourteen months, Anslinger characterized it very differently from the medical research he had gathered and kept in his own files, as evidenced by the following etymological argument.

This argument—the origin of the word *hashish*—pervaded Anslinger's marijuana statements throughout his tenure and prepared his listeners to accept his proposals. To a nation uneducated in narcotic drugs, Anslinger could use the terms "marihuana" and "hashish" interchangeably, despite the drugs' differences. Though both come from the hemp plant, marijuana is the plant's dried flowers and leaves, and hashish is the product of a longer drying process of hemp flowers and leaves, resulting in a sticky resin that is pressed into blocks, and is a much more potent narcotic.

According to the commissioner, the Persian military in 1090 A.D. was known as the "Hashishans" because its leaders would solicit soldiers by giving them hashish. Once under its influence, with no concern for their own safety, the docile soldiers would perform violent and bloody acts upon command and often carry out "secret murders." Anslinger frequently made the connection between the English word *assassin* which, he said, appropriately derives from the Arabic word "Hashishan" and which he would change to "people who use marihuana." This argument was offered to convince listeners that for centuries marijuana use contributed to violent crime.[27]

However, what Anslinger's listeners probably did not know was that Anslinger told only half of the assassin story. The *Oxford English Dictionary,* an etymological dictionary, provides a detailed explanation of the origin of the word *assassin:* Hassan-ben-Sabbah took a group of people to a garden that exceeded their most voluptuous imaginings, where he exhausted all of their strengths and desires and told them they must ignore divine laws and blindly obey him, the Old Man of the Mountain. The people felt that Hassan had given them paradise, and to ensure their stay there, both in this life and beyond, they needed only to put their lives in his hands and to follow his orders without question. This included murdering and stealing and even committing suicide. Apparently fond of demonstrating his power over his followers, Hassan ordered one person to jump from a window and another to behead himself, and both obeyed immediately. When Henry of Champagne visited Hassan, he demonstrated his power by signaling to two guards to jump from the tower to crash at Henry's feet. They did so, and Hassan assured Henry that, at Hassan's signal, all of the others would follow.

Because Hassan reportedly gave them hashish, a word derived from his own name, to inebriate them, the word *hashishan* came to describe them as thieves and murderers. (*The Murderers* was later a title of one of Anslinger's books.) Because Anslinger used these words interchangeably, this story enabled him to connect hashish or marijuana with violent crime, to demonstrate that the use of the drug incited the user to violence. The parts he distorted—portraying a band of ruthless killers who used hashish as a prerequisite to performing murder and omitting Hassan from the story—painted a very different picture. Hashish reportedly played a part in the ritual that Hassan instructed the people to follow. However, the causal argument breaks down because the murders were not caused solely by use of the drug; the murders were caused by the entire ritual that happened to involve the use of hashish—and not marijuana—and the fact that Hassan promised this paradise to those who without question would follow his orders.[28] This argument is post hoc: it claims that because the violence occurred *after* the people used hashish, the violence occurred *because of* the hashish. Yet Anslinger distorted the facts and used this story throughout his thirty-two-year appointment to convince people of the connection between marijuana and crime by placing marijuana in a historical context of violence.

Anslinger was not alone in his quest to maintain his bureau and to fight the perceived marijuana menace. Immediately following the Doughton Bill announcement, groups that had an interest in the FBN supported its continuance, using the media and conducting letter-writing campaigns. Because they were so vocal, their effort appeared greater than it was. Among those who assembled to defend Anslinger and his bureau were drug manufacturers and wholesalers, druggists, physicians, and temperance groups.[29]

Numerous letters crossed the country in support of Commissioner Anslinger and the FBN. The campaign worked: the October 1, 1936, issue of the *N.A.R.D. Journal* reported, "The bill was scheduled at first to be quietly pushed through the Congressional mill without hearing. . . . A flood of protests resulted and as a final result, the bill failed to get the approval of the Committee."[30]

Although Anslinger's public statements prior to the introduction of the Doughton Bill contained almost no references to marijuana, after January 1936 his speeches became filled with causal claims connecting marijuana use to crime and insanity: he sought occasions to speak on the subject, volunteered marijuana testimony before Congress, and began to collect files of marijuana-related horror stories. In August 1936, Anslinger received a request from J. Edgar Hoover to speak "on the enforcement of narcotic laws, before the third session of the FBI National Police Academy." In his memo to Assistant Treasury Secretary H. E. Gaston requesting permission to accept the invitation, Anslinger wrote, "This would afford an excellent opportunity for me to talk to the Police of the various cities throughout the country and I feel that it would be quite beneficial, particularly because of the present menace of MARIHUANA."[31]

By the end of 1936, Anslinger's verbosity on the topic of marijuana backfired: despite his support for the Uniform State Narcotic Drugs Act, marijuana jurisdiction would shift from the state to the federal level. When the public's desire for a federal marijuana law became too great to ignore, Anslinger agreed to help draft the legislation. On January 14, 1937, fourteen people met at the Conference on *Cannabis sativa* L. in Washington, DC, to define and analyze the parts of the marijuana plant as well as the drug's uses and effects. The group comprised doctors, researchers, lawyers, and government officials, including Anslinger.

Responding to a question about "undesirable characteristics" that might result from smoking marijuana, Anslinger said that the FBN had "a lot of cases showing that it certainly develops undesirable characteristics." He then read the case of a fifteen-year-old boy who went crazy on an Ohio playground after the playground supervisors sold him some marijuana. Responding to Anslinger, S. G. Tipton of the general counsel's office said, "Have you a lot of cases on this—horror stories—that's what we want."[32]

Anslinger then asked Dr. Carl Voegtlin, chief of the Pharmacology Division of the National Institutes of Health, whether cannabis "leads to insanity." Voegtlin replied, "I think that it is an established fact that prolonged use leads to insanity in certain cases, depending on the amount taken, of course. Many people take it and do not go insane, but many do." Although Voegtlin had qualified his response, Tipton repeated, "If it leads to insanity, and we have a lot of horror stories, we can build it up."[33]

Almost immediately, the committee got what it wanted. A barrage of horror stories was printed in 1937, including an article the commissioner co-authored with Courtney Ryley Cooper titled, "Marijuana: Assassin of Youth," published in *American Magazine*.[34] Articles with titles such as "Youth Gone Loco," "One More Peril for Youth," "Marijuana—A New Menace to U.S.," and "Danger" deliberately exaggerated marijuana's effects on young people.[35]

In a charge to police officers, Anslinger and Cooper wrote, "It would be well for law enforcement officers everywhere to search for marijuana behind cases of criminal and sex assault," adding that at least two dozen recent murders or sexual assaults involved marijuana.[36]

During the 1937 House Committee on Ways and Means' public marijuana hearings, the AMA spoke out in response to the FBN's distorted connections between marijuana and crime. William C. Woodward, legislative counsel for the AMA, was the only witness at the Marihuana Tax Act (MTA) hearings who questioned the bureau's evidence. On May 4, 1937, Woodward testified that although the AMA was concerned about the alleged connections, it was also concerned about the obvious lack of knowledgeable witnesses at the proceedings.

Woodward had conducted sufficient research to substantiate his position. To rebut the FBN's claim that marijuana caused crime, the doctor previously contacted the Bureau of Prisons, the Children's Bureau, the Indian Bureau, the Office of Education, the Public Health Service, and the Division of Mental Hygiene. Not only was Woodward dismayed that no experts from these bureaus were called to testify, but his own informal inquiries with those bureaus showed first that none of them had investigated the marijuana-and-crime connection, and second that none of them knew that it was an area of concern. In no way minimizing the potential seriousness of marijuana abuse, Woodward presented the AMA's opinion that marijuana should be controlled at the state level and that a federal tax was a nuisance and would not result in curbing marijuana traffic.

Despite Woodward's introducing himself with regret—that for the first time as the AMA's representative, he found it necessary to take a stand in opposition to the government—his apology did nothing to alleviate the hostile reaction of Doughton and the committee members. Nor did it prevent them from accusing him of being "simply piqued because you were not consulted in the drafting of the bill." Dr. Woodward replied, "Not in the least. I have drafted too many bills to be peeved about that."[37]

Through thirty-four pages of testimony, the committee members harassed and badgered the doctor, accusing him of not cooperating with the committee and finally telling him that they simply were not impressed with his testimony. No other AMA members testified before either the House or the Senate during these hearings.

Previously, the committee had welcomed Anslinger's testimony, replete with gory stories linking marijuana to rape, murder, incest, and insanity. He also provided what Woodward called "hearsay testimony" from various prison officials and doctors. However, unlike the hostile interruptions Woodward endured during his testimony, through Anslinger's thirty-five pages of testimony, the committee interrupted him only once. Further, the committee gave the commissioner every opportunity to embellish his examples and elicit further emotional responses from his audience. Anslinger outlined at length the route of many marijuana users who

> fly into a delirious rage after its administration, during which they are temporarily, at least, irresponsible and liable to commit violent crimes. The prolonged use of the narcotic is said to produce mental deterioration. . . . Among some people the dreams produced are usually of an erotic character. . . . Then follows errors of sense, false convictions and the predominance of extravagant ideas where all sense of value seems to disappear.
>
> The deleterious, even vicious, qualities of the drug render it highly dangerous to the mind and body upon which it operates to destroy the will, cause one to lose the power of connected thought, producing imaginary delectable situations and gradually weakening the physical powers. Its use frequently leads to insanity.
>
> I have a statement here, giving an outline of cases reported to the Bureau or in the press, wherein the use of marijuana is connected with revolting crimes.[38]

To politicians, Anslinger was their ticket to reelection, especially when the legislation was supposedly saving America's most valued resource—its youth. Therefore, the legislators were motivated to help him in his authoritative appeal—which contained very little substantiation and no scientific evidence.

Anslinger's files also contained several passionate letters from physicians who acted as witnesses in criminal cases involving marijuana throughout the country, attesting to the lack of evidence of a connection among marijuana, crime, and insanity.[39] These people showed that Anslinger's arguments were fallacious. Anslinger responded by treating them as the enemy.

Yet evidence in Anslinger's files shows that Anslinger had previously agreed with Woodward and others. As early as March 20, 1933, in an address titled "Peddling of Narcotic Drugs," Anslinger emphatically stated,

> There is probably no more absurd fallacy extant than the notion that murders are committed and robberies and holdups carried out by men stimulated by narcotic drugs to make the[m] incapable of fear. This may occasionally happen, but the immediate effect of a narcotic drug is usually to soothe the abnormal impulses, and the ultimate effect is to create a state of idleness and dependency.[40]

Nevertheless, on October 1, 1937, the MTA went into effect.[41] Because some legitimate use for the hemp plant existed, the act called for a transfer tax of $1 an ounce for persons registered with the government as handlers and $100 an ounce for those not registered. Possession by people not involved in the named industries was punishable with a $2,000 fine, five years imprisonment, or both; those penalties were enforced by the FBN. Thus, with the MTA, marijuana enforcement policy shifted from the state level to the federal and granted the FBN complete authority over it. Although the MTA did not prohibit the therapeutic use of marijuana, it made therapeutic use so difficult and frightening for physicians and patients that most physicians no longer prescribed it. Those physicians who did register for the $1 tax stamp were subsequently required to report to the FBN the patients' names and addresses, amounts of marijuana prescribed, and the illnesses for which it was prescribed. Any physician who for any reason did not immediately file the required paperwork put himself and his patients at risk for the MTA's high monetary penalty and long prison term. Thus, using what Congress passed as a revenue measure, the FBN started a trend of deterring marijuana use through punitive measures—a trend that, although lessening, continues to the present.

THE BOGGS ACT OF 1951

During much of the 1940s, World War II and its aftereffects occupied the attention of most people in the United States. Anslinger and the Federal Bureau of Narcotics maintained a low profile during the war years. Anslinger did not campaign for any drug policy reform during the 1940s. Nor did he request budget increases, as limited appropriations resulted in the demise of many federal agencies and programs. His annual report on December 14, 1942, to the congressional Appropriations Committee for fiscal year 1944 was only one year and one week after the bombing of Pearl Harbor, and at that point the United States was firmly entrenched in the war. Anslinger carefully followed Chairman Louis Ludlow's dictum as he opened the appropriations hearings on December 10, 1942. Ludlow said,

> It would be a refreshing novelty, most encouraging to the country, if some of the officials when they come before us would vary the precedents a little by telling us not so much of the appropriations they think they must have but the appropriations which they think they can manage to do without. The temper of the country was never so set against waste as it is today. Let there be no mistake about that. The people of this country, harassed by Government exactions due to the war and bled white by taxation, with more taxes to come, are expecting the newly elected Members of the Seventy-eighth Congress to set a new record of watchful care for the taxpayer's interest and to pursue rigid economy in all appropriation matters not directly connected with national defense.[42]

Because the FBN's budget was considered nondefense, Anslinger took Ludlow seriously and appeared before the subcommittee with an estimate of his 1944 budget showing a decrease of $89,060, or 7 percent, from the 1943 budget.[43] However, while Anslinger appeared satisfied that he had demonstrated his and the bureau's awareness of the war effort, Representative Ludlow probably surprised Anslinger by bluntly asserting that the FBN's "activities are not especially related to the war effort; they are related to national welfare." Anslinger's response was brief: "They are related to both. Our work with respect to critical and strategic materials all ties into the war effort."[44]

That the FBN was involved in the war effort even before the United States became involved was evident from an April 9, 1941, memo in which Anslinger outlined a case at Fort Eustis, Norfolk, Virginia, involving a draftee, who was also a New York racketeer and pimp, who set up "houses of prostitution for the convenience of soldiers and sailors" in Norfolk. The women, whom the individual had brought from

New York, "had been throwing marihuana parties with the draftees."[45] At this appropriations hearing, Anslinger did not offer this information.

However, the next year, Anslinger was prepared, even putting the topic on the table himself. In the middle of his testimony, Anslinger shifted the topic to what he called the "big problem in the past year": he stated that while the FBN investigated some morphine theft from Army medical depots, the real problem was the use of marijuana in Army camps. This problem had involved "20,000 man-hours," with "3,000 investigations pending," and "25 agents whose time is devoted exclusively to such investigations"; in fact, the commissioner stated, "the situation is so serious that I have taken up with the Provost Marshal General the necessity of having Army personnel assist our forces, because we just cannot handle all these investigations that are coming in now," indicating that the FBN was in fact involved in the war effort and that the need for additional agents was greater than ever.[46]

In his testimony, Anslinger cited three areas in which marijuana was a concern for the military: first, "Army personnel deliberately are caught with marihauna in order to endeavor to obtain discharges from the Army." Second, some young men found marijuana useful in their efforts to evade the draft. Third, the armed services' need for hemp to make rope and other materials necessitated large acreages of licit marijuana; however, because they were difficult for the FBN to supervise, often even these registered acres were the source of the illicit supply of marijuana finding its way to the Army. Thus, once again, the mild narcotic marijuana came to the bureau's rescue, enabling Anslinger to demonstrate the FBN's contributions to the war effort.[47]

After World War II, new fears gripped the nation. The media, with no war to cover, devoted more time and attention to drugs. In addition, the media capitalized on the national fear of Communism, the Mafia, organized crime, the atomic bomb, and, to some extent, drugs. Fear was so strong that civilization itself seemed threatened.[48]

The United States was unaccustomed to living with the lingering fear that remained after World War II. For the nation as a whole, this fear was critically different from other earlier, passing fears that could be fought and won or talked through and settled. Those were fears with origins outside of the United States. These new fears were internal, something intangible "oozing" through life in the United States. These fears were brought on by a combination of New Dealers and intellectuals who, many thought, were trying to socialize the United States. Added to this concern was the fear of Communists—working within the United States without conventional weapons of war—silently infiltrating American culture.

The public was well aware of this internal subversion through media coverage of the House Un-American Activities Committee, particularly the 1948 and 1949 trials of Alger Hiss. The results increased people's fear of Communists within the United States, the bitterness toward them, and the paranoia that a neighbor could be one of them. At a press conference in January 1950, President Harry S Truman "talked grimly of the 'great wave of hysteria' that was building. It would subside, Truman added staunchly. We had gone through this sort of thing before and 'the country did not go to hell, and it isn't going to now.'"[49]

President Truman's impatience with the country's hysteria set the tone for Commissioner Anslinger's public statements in the early 1950s. Anslinger held a presidentially appointed position, and following the president's lead was simply a smart and politic thing to do. So for a time, Anslinger shifted dramatically to a new role as ameliorator and calmer of the public's fears about illicit narcotics use, particularly among young people.

Fortunately for Anslinger, during the 1930s he had successfully established his arguments; then others had picked them up and carried them forward into the 1950s. Therefore, Anslinger could follow the president's directive to quash the hysteria because a small chorus of congresspersons was keeping alive his old arguments about the relationship between drug users, insanity, and violence. Through the 1950s Anslinger repeatedly argued five claims: first, longer prison sentences would deter narcotics peddlers and would cause "the traffic [to] just about melt away";[50] second, to stop narcotics traffic, the FBN needed more agents; third, narcotics education would have more deleterious effects in the form of more experimentation among young people and therefore should not be undertaken; fourth, teenage drug addicts did not come from good homes; and fifth, the problem was not as widespread as people believed and was, instead, the result of hysteria and newspaper publicity. Through the 1950s Anslinger remained steadfast in these claims. However, through that decade Anslinger did not resort to the horror and gory tales of the 1930s. Others were doing that for him.

On February 27, 1950, marijuana returned to the attention of Congress. New York Representative Jacob K. Javits reconnected marijuana to crime, saying,

> Mr. Speaker, an alarming number of crimes in various sections of the United States have been committed by young people who have been under the influence of marijuana. Many of these crimes have been committed by teen-age youngsters who, while under the influence of this vicious drug, have committed murder, robbery, and other equally shocking crimes.[51]

He cited an example of "an innocent citizen" who was robbed and murdered by two girls, ages eighteen and fifteen, one of which admitted she was under the influence of marijuana.

Javits urged citizens to begin to take control of the problem: "Do not wait for the other fellow to do it; do it yourself and do it now," and he ended by urging that the "far-reaching medium of radio" take part in not only reporting the activities of criminals but in more actively "showing the criminal in his true light—as a coward, a menace to decent citizens, a broken wreck of a human being, often a marijuana addict or user of other drugs." Such media coverage, Javits believed, would "bring an end to this evil."[52]

The subsequent media coverage did not end the evil, and for good reason: addiction among young people was increasing. A November 5, 1950, *New York Times* article, "Youth Narcotic Use Growing," quoted Victor Vogel of the U.S. Public Health Hospital in Lexington, Kentucky, saying, "It looks like a definite effort on the part of peddlers to contact teen-agers."[53] In the February 2, 1951, issue of *The New York Times,* Kenneth W. Chapman, director of hospital training for the Public Health Service, offered statistics about "youthful addicts" admitted to the hospital: "In 1948, three per cent of the admissions were of patients under twenty-one. In the first six months of 1950, eighteen per cent were in that age group."[54]

The language and detail of each narcotics item in the media escalated the scare. While the above two articles were just two column inches each, a March 2, 1951, *New York Times* article was twice as long and quoted a Narcotics Bureau report that specifically mentioned the peddlers' trick of giving free narcotics to teenagers "to start them on the road to addiction."[55]

Early in 1951, two important hearings brought Anslinger back into the limelight: the Senate Special Committee to Investigate Organized Crime in Interstate Commerce and the hearings on House Resolution 3490, A Bill to Amend the Penalty Provisions Applicable to Persons Convicted of Violating Certain Narcotic Laws, and for Other Purposes, also known as the Boggs Act. These two hearings provided Anslinger with opportunities to promote his claims, in particular his lifelong belief that the primary solution to drug problems was longer prison sentences. However, Anslinger tempered his statements with cautions to avoid hysteria (see Photos 3.3 and 3.4).

On March 27, 1951, Anslinger appeared before the Senate Special Committee to Investigate Organized Crime in Interstate Commerce. Chaired by Tennessee Senator Estes Kefauver, it was known as the Kefauver Committee.[56] In his opening remarks before the commit-

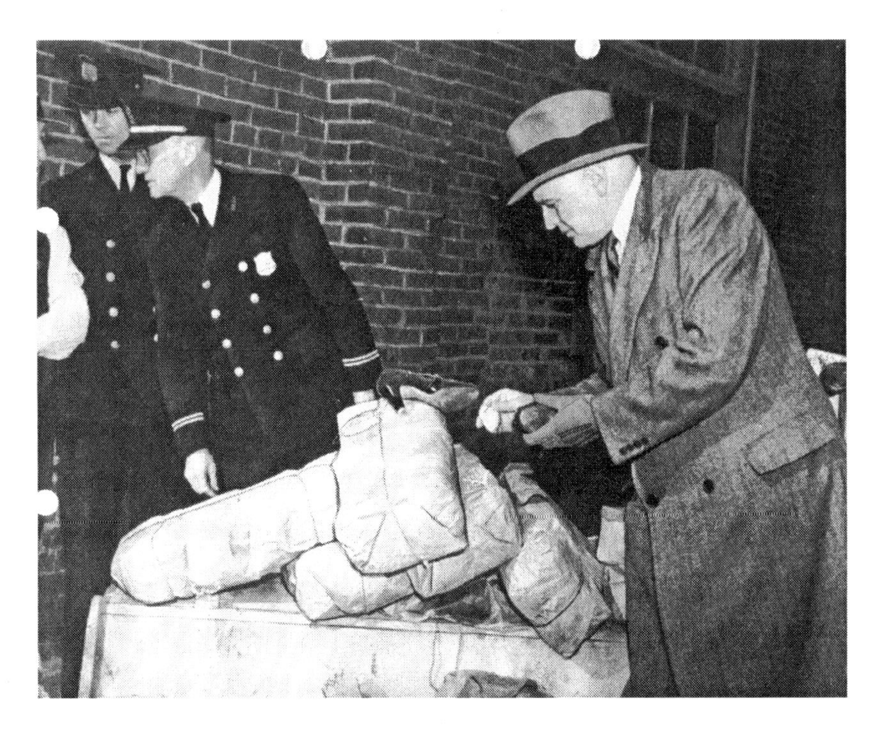

PHOTO 3.3. Harry J. Anslinger examines evidence. (*Source:* Harry J. Anslinger Collection, courtesy of Historical Collections and Labor Archives, Special Collections Library, The Pennsylvania State University [Folder #3])

PHOTO 3.4. Harry J. Anslinger speaks at a congressional hearing on drug control. (*Source:* Harry J. Anslinger Collection, courtesy of Historical Collections and Labor Archives, Special Collections Library, The Pennsylvania State University [Folder #3])

tee, Anslinger attempted to discount the reports of increased drug abuse, saying, "The facts of the recent increase in drug addiction are somewhat obscured by hysteria and good news copy," a surprising remark in general because Anslinger's rapport with the media was quite good and in particular because his own previous statements had generated their own share of hysteria.[57]

Keeping hysteria out of his own testimony, Anslinger offered statistics from the federal drug hospital in Lexington, testifying that a much larger percentage of young people were reporting for treatment of drug addiction; in fact, he testified, federal drug hospital admissions of people under age twenty-one increased from 3 percent in 1946 to 18 percent in 1951. Anslinger blamed this increase on "a wave of juvenile delinquency" and, paraphrasing his "good home" theory, stated, "The type of young hoodlum addicts we are encountering today are not the type who can be deterred by copy-book maxims or reached by an educational program."[58] This statement reinforced his position that education did little to prevent drug experimentation.

Anslinger's testimony alleviated some of the fears, first by assuring people that the drug problem was not as widespread as the media portrayed it, and second by showing that the FBN had uncovered the existence of the Mafia in the United States and had effected the convictions and deportations of a number of its drug kingpins.

At the end of his statement, Anslinger submitted six proposals for dealing "more drastically and realistically" with narcotics traffic: first, longer prison sentences; second, more FBN agents; third, more narcotics squads at state and local levels of law enforcement; fourth, a witness protection program; fifth, a list of interstate racketeers to be disseminated to state and local enforcement agencies; sixth, compulsory confinement of addicts until "they are pronounced cured by medical authorities."[59]

The Kefauver Committee appreciated Anslinger's presence and expertise at the hearings. In Estes Kefauver's book on the subject, *Crime in America,* he wrote, "Our greatest help in tracking down the trail of the Mafia came from the Federal Bureau of Narcotics." The FBN's connection to the Mafia, Kefauver explained, stemmed from the "Mafia's dominance in the dope trade."[60] The accolades that Anslinger and the bureau received as a result of his testimony contributed not only to the continued existence of the FBN but also to its future expansion.

At the end of its investigation, the Kefauver Committee made twenty-two recommendations, one of which responded to Anslinger's testimony: "general increases in personnel of seriously understaffed federal law enforcement agencies and elimination of inequities in the salaries of these federal law enforcement officers, many of whom, we noted, 'are woefully underpaid' for the difficult and dangerous duties they perform."[61]

In addition, the recommendations supported a number of the commissioner's requests. The committee wrote,

> Penalties against the illegal sale, distribution, and smuggling of narcotic drugs should be increased substantially. There is an alarming rise in the use of narcotics, particularly among teen-

> agers, who begin with marijuana and gradually become hopeless addicts to heroin and cocaine. The average prison sentence meted out to narcotics traffickers is eighteen months, and, as Narcotics Commissioner Anslinger told us, "short sentences do not deter."[62]

The committee recommended that Congress enact legislation requiring a mandatory five-year prison sentence to anyone convicted of selling narcotics to a minor, a recommendation that in 1956 would become reality but not without further prodding from the commissioner.

One month after his appearance before the Kefauver Committee, Anslinger testified in support of HR 3490, the Boggs Act. Compared to others, Anslinger's testimony was brief, probably because he was the second FBN representative to testify. George W. Cunningham, FBN deputy commissioner, testified before Anslinger on the bureau's behalf, as Anslinger had a previous commitment to prepare for a meeting of the United Nations' Commission on Narcotic Drugs. Their testimony focused primarily on two claims: first, that public hysteria was making the drug problem seem larger than it was, and second, that the single most effective deterrent to violation of drug laws was long, mandatory prison sentences, a major provision in the Boggs Act.

Because Anslinger testified after Cunningham, Anslinger's testimony was somewhat less factual and more philosophical, and eerily reminiscent of his Progressive upbringing. Anslinger and Boggs talked at length about "aliens" in the drug business. Anslinger reported that the FBN had deported hundreds of aliens, more than any other agency, "because we can deport immediately upon conviction. . . . We do not have to have a second conviction on that."[63] However, Anslinger pointed to a "defect" in the legislation regarding alien peddlers who were also addicts. This defect allowed for hospitalization of alien addict peddlers rather than immediate deportation. Anslinger said, "I think we should even get rid of that defect in the legislation, because I think they have sympathy for an alien who is an addict."[64]

Anslinger's most consolidated media campaign for the Boggs Act occurred during June 1951, one month prior to the House vote. The June 18, 1951, issue of *The New York Times* contained a front-page article with the subheadline, "Juvenile Addiction at Epidemic Level in Nine Cities, Federal Commissioner Declares." Another subheadline read, "Stricter Laws Urged," reiterating Anslinger's arguments that higher fines and longer jail sentences deterred would-be criminals. The article began with the declaration that addiction among youth in nine major cities had reached "epidemic proportions." It then quoted Anslinger, who "blames [the problem on] . . . lack of parental control and the 'social disintegration' of the large urban centers."[65] In this release, Anslinger connected narcotics and crime, although not as the effect of any drug, but as the young drug users' source of money to purchase their drug supplies. This crime wave on the part of juveniles, according to Anslinger, was "bringing the problem into the open."

Anslinger claimed that "teen-age youths still in high school" were the newest victims, although only in New York City.[66] His bureau could successfully respond to this epidemic before high school students in other cities were affected only if the FBN was equipped with the proper "tools." Anslinger then attempted to design for his bureau a much more powerful tool kit by requesting "compulsory prison terms of five, ten and twenty years for second and third offenders." The article stated, "Congress now is considering such legislation."[67]

Anslinger maintained his calm, factual approach in an interview in the June 29, 1951, issue of *U.S. News and World Report*. About drug prevention education, Anslinger said, "They say, educate them. But what education can you give children who are not in school? In a weak mind? Education on narcotics places ideas. I don't think it is a wise thing." Later in that interview, when asked about the "relative use" of marijuana, cocaine, and heroin, Anslinger responded, "They start on marijuana, then graduate to heroin." The *U.S. News* reporter asked, "Is it [marijuana] habit forming? Is it as dangerous as other narcotics?" Anslinger responded, "It is habit forming but not addiction forming. It is dangerous because it leads to a desire for a greater kick, from narcotics that do make addicts."[68]

Further, Anslinger used this press opportunity to point out the need for longer jail terms for peddlers. When Anslinger referred to the issue of teenage drug abuse as a problem that "can be stopped," the reporter asked, "How?" Anslinger responded,

> I think the situation in St. Louis probably is cured by the fact that Federal Judge Roy W. Harper gave a peddler there eighteen years. There is a general exodus. . . . We have 180 agents. It's

> like using blotting paper on the ocean. But we catch them—the smugglers, the syndicates, the pushers, and wholesalers, and the users. We can catch them. But we can't keep them in. They serve about 16 months. We put one crowd in jail, then start on another one. By the time we get the second one, the first is out working again. So it's just a merry-go-round.

Anslinger then suggested that Congress could help by passing bills

> to increase the penalties to a minimum of two, five, and ten years for first, second, and third offenses. [Illinois] Senator [Everett] Dirksen is introducing a bill making it life for the sale of narcotics to minors. We are going to support that.

He commended four states—Tennessee, West Virginia, New Jersey, and Maryland—for increasing penalties.[69]

The reporter from *U.S. News and World Report* asked further, "Do parents generally need to worry about this increasing use of drugs among young people?" Anslinger reiterated his feeling that recent reports about drug abuse among young people were exaggerated: "Not if they look after their children properly. We don't find addicts among children from good homes. People get a bit hysterical about reports of narcotics sales around school children."[70]

Despite Anslinger's claim that young people were not in any particular danger, articles contradicting that belief continued to appear. Anslinger's attempts to contain his own former claim, now being repeated by others, failed. The February 26, 1951, issue of *Time* contained a brief article, "Youth: High and Light," with a report of a man who found his fifteen-year-old son shooting heroin. The article stated, "Weeping, the boy confessed that he had used drugs for a year—first marijuana on a dare from a schoolmate, then the virulent morphine derivative, heroin." To purchase their drugs, "some boys become thieves and holdup artists; many a teen-age girl has turned to prostitution." Typically, the article stated, "Penalties for the vicious crime of dope-peddling are too lenient (maximum: ten years) to deter many from the hugely profitable trade." However, the article ended with a list of approaches that various cities and states were taking to combat the problem. Among them, New York and Chicago were increasing their narcotics police forces; New York schoolteachers were in training to recognize symptoms of addiction; and "even more to the point," the Illinois legislature was considering a bill "which could send dope peddlers to prison for life for selling narcotics to a minor."[71]

The June 27, 1951, issue of *Pathfinder* also contained a story, "Drug Peddling, the Dirtiest Crime," about New York City junior high school students who wrote compositions on the topic "What I Know About Narcotics." This kind of article no doubt contributed to the hysteria that Truman and Anslinger tried to calm. The students admitted that they "were smoking 'reefers,' 'snorting' heroin, and 'going on the needle' within the schools themselves—in the lunchroom or down in the boiler room or up on the roof." The students wrote that at first the price is cheap; "marijuana cigarettes can be had for 75 cents apiece." One "boy bookie" explained, "Naturally, if they continued the habit, the price would go up to $3, $3.50." The solution, according to the article, was more stringent penalties such as the House Ways and Means Committee's vote for "minimum penalties of two, five, and ten years imprisonment for all narcotics vendors."[72]

A brief July 14, 1951, *Science News Letter* article titled "Child 'Dope' Addicts" agreed with Anslinger that the number of young addicts was exaggerated. The article cited information from the U.S. Public Health Service Hospital at Lexington and stated, "Addiction of children and adolescents to narcotic drugs, from marihuana to heroin and morphine, apparently is not so widespread throughout the whole nation as it seems from accounts of the current investigations."[73]

In an October 1951 article, "Control of the Traffic in Narcotic Drugs," that appeared in *The Merck Report,* Anslinger outlined the history of narcotics legislation in the United States to enable his audience to see the necessity of the proposed Boggs Act.[74] Passage of the Boggs Act would increase the ten-year maximum sentence for drug offenders to a two-to-five-year sentence for first offenders, a mandatory five-to-ten-year sentence for second offenders, and a mandatory twenty-year sentence for third offenders; second and third offenders had no chance of probation or parole, and all offenders could be fined up to $2,000. About HR 3490, Anslinger wrote,

> If enacted into law, it should prove a most important aid to narcotic law enforcement by removing the persistent peddler from

> his illicit activity for a longer period of time, while acting as a strong deterrent to those who would normally seek to become his successor in such traffic.[75]

The article featured a close-up of a marijuana leaf and a picture of bushy marijuana plants growing wild "as a wayside weed," as a poignant reminder of the drug's availability and of the need for more enforcement officers.

Many congresspersons treated the problem as large and threatening, despite Gallup poll results that showed that those surveyed did not consider drugs a problem in the United States.[76] However, some members of the House of Representatives did oppose HR 3490. According to a July 17, 1951, article in *The New York Times,* "The House of Representatives approved by voice vote today a bill providing severe penalties for violation of Federal narcotics and marijuana laws and sharply restricting judicial discretion in applying them."[77] The article added, "Opposition in the House was substantial," specifically citing New York Democrat Emanuel Celler, Pennsylvania Republican Richard M. Simpson, and Pennsylvania Democrat Herman P. Eberharter.

These representatives' main argument "was that the bill, in seeking to wipe out illicit traffic in narcotics, used 'a blunderbuss loaded with scattershot, that would claim innocent victims as well as the guilty.'" In other words, the representatives feared that leaving judges with no flexibility in second- and third-offense cases "might do an injustice to those who fell victim to the use of drugs." A second argument dealt with the effects of putting judges in a "straight-jacket" that forced them to impose heavy penalties. The opposition foresaw grand juries that "refuse to indict" and petit juries that "refuse to convict." Finally, "others in opposition" suggested that current legislation was sufficient but that the courts needed to impose and enforce it more stringently.[78]

The opposition did not even slightly slow down the Boggs Act's passage.[79] In fact, on October 20, 1951, HR 3490 passed the Senate without a word of debate or objection. Introducing it that day, the Senate presiding officer said, "Is there objection to the present consideration of the bill? The Chair hears none; and, without objection—."[80]At that moment, Maryland Senator Herbert O'Conor interrupted, saying that such an important bill should not pass without a word of discussion, and he proceeded to reiterate the need for the provisions of the bill.

The bill was then read a third time and passed with no objection. At that moment, Senator Kefauver requested the floor long enough to make a statement that, similar to O'Conor's, was fairly self-gratifying. Both senators spoke about their work on the Senate Crime Committee. Finally, Kefauver quoted Anslinger, referring to the increased penalties, saying, "I think it would just about dry up the traffic."[81]

The New York Times reported that, while signing the Boggs Act on November 2, 1951, President Truman remarked that, while he was aware that there was "some objection" to the bill's "limitations on Federal courts in sentencing offenders," he felt that "the effects of drug addiction upon the individual, the family and the community as a whole are only too self-evident."[82]

Subsequently, the January 23, 1952, issue of *Pathfinder* magazine reported on an FBN nationwide drug raid. Reporter John M. Conly wrote,

> The Bureau is not sentimental about any drug addicts, juvenile or not. Addiction among "legitimates" (their word for people not describable as criminal or semi-criminal) is rare. The drug addict is someone already at odds with society. The Bureau's main aim is not to "save" him. It is to save society from him. Drugs are a tremendous stimulant to gang-formation, and addiction makes a bad criminal worse. Hence the Bureau believes in being tough. It is happy about the Boggs law. . . . And it is happy about Baltimore. Alone among large cities, Baltimore yielded no arrests in the big sweep. Likeliest reason: Maryland has had since last June a law just like the Boggs law. . . . With the help of 42 state legislatures, Harry Anslinger thinks, the whole United States might be made too hot for drug-peddlers.[83]

With the Boggs Act, Anslinger was successful in securing legislative support for his most pervasive argument, to show that prison sentences needed to be longer and fines higher. Yet while Anslinger was successful in increasing sentences for narcotics violations and was correct in his statement that the FBN was responsible for approximately 10 percent of the prison population, the effectiveness of such legislation should not be measured only in these terms. Taken from Anslinger's yearly testimony before the Congressional appropriations subcommittees, the following figures show that, in 1950, 4,530 people were convicted for drug violations. This conviction figure continued to drop through the 1950s—4,079 in 1951; 3,236 in 1952; 2,650 in 1953; 2,197 in 1954; and 2,134 in 1955 (Figure 3.1). These numbers suggest that those congresspersons who opposed the Boggs Act were correct in their reservations about courts becoming less inclined to convict when conviction meant a mandatory minimum sentence with no chance for parole or suspension.[84] This yearly decline in convictions could also support Anslinger's deterrence theory and David Musto's theory of intolerance for the decade of the 1950s.[85] Unfortunately, the numbers that would clarify this issue—namely, total number of arrests for narcotics violations in comparison to total number of convictions—are not readily available.

However, while the conviction numbers dropped, another factor that needs to be considered in terms of successful legislation is the

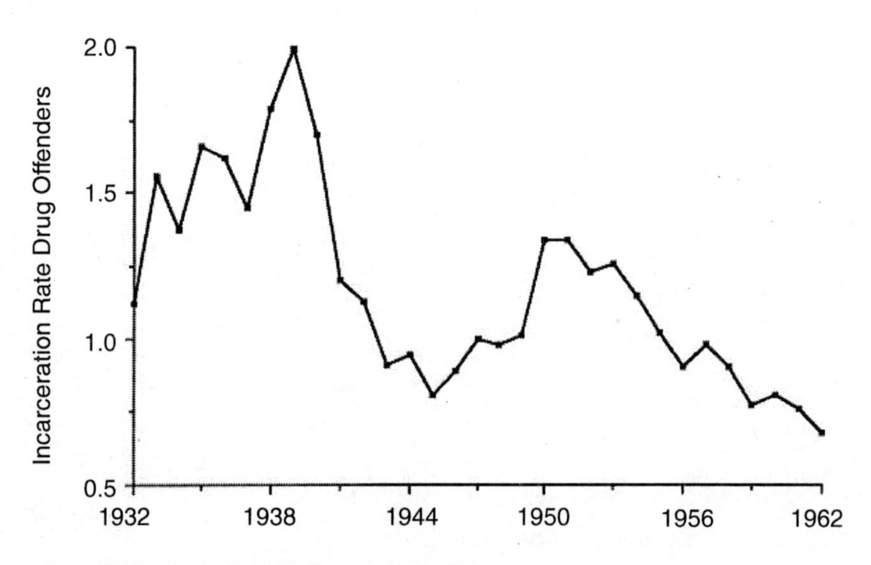

FIGURE 3.1. Offenders sent to prisons for drug offenses per 100,000 people from 1932-1962

number of people convicted who actually served time. In that regard Anslinger was successful. Before the Boggs Act was passed in 1951, of the 4,530 persons convicted of narcotics violations in 1950, only 2,029 or 44 percent served time.[86] In 1951, of the 4,079 persons convicted, only 2,063 or 50 percent served time, and in 1952, of the 3,236 persons convicted, 1,932 or 59 percent served time. This percentage rose significantly with the passage of the Boggs Act in November 1951: in 1953, of the 2,650 persons convicted, 2,016 or 76 percent served time; in 1954, of the 2,197 persons convicted, 1,875 or 85 percent served time, and in 1955, of the 2,134 persons convicted, 1,694 or 79 percent served time (see Figure 3.2). Anslinger would continue these arguments throughout the next decade, the remainder of his tenure as commissioner.

On November 3, 1951, when *The New York Times* reported President Truman's signing of the Boggs Act the previous day, the article ended with a prophetic note. The national legislative committee of the Veterans of Foreign Wars voted the same day to "urge stronger state narcotics laws, with stiffer penalties for offenders," including "the death penalty for persons selling narcotics to teen-agers."[87] Narcotics traffic did not "dry up" as a result of the Boggs Act. The quest for even more stringent penalties would continue.

FIGURE 3.2. Convictees for drug offenses per 100,000 people from 1932-1956

URLs FOR SELECTED PRIMARY DOCUMENTS

The Traffic in Narcotics by H. J. Anslinger
<http://www.druglibrary.org/schaffer/people/anslinger/traffic/default.htm>

Statements of H. J. Anslinger
<http://www.druglibrary.org/schaffer/hemp/taxact/anslng1.htm>
<http://www.druglibrary.org/schaffer/hemp/taxact/t3.htm>

"Narcotics Bureau Is Urged by Tuttle," March 9, 1930, *The New York Times*
<http://www.druglibrary.org/schaffer/history/e1930/tuttle.htm>

Conference on *Cannabis sativa* L., January 14, 1937
<http://www.druglibrary.org/schaffer/hemp/taxact/canncon.htm>

Report on Marijuana Investigation, Summer of 1937
<http://www.druglibrary.org/schaffer/hemp/taxact/summer.htm>

Marihuana Tax Act of 1937
<http://www.classicpharm.com/taxact.html>
<http://www.druglibrary.org/schaffer/hemp/taxact/t1.htm>
<http://www.druglibrary.org/schaffer/hemp/taxact/mjtaxact.htm>

Taxation of Marihuana, May 4, 1937
<http://www.druglibrary.org/schaffer/hemp/taxact/woodward.htm>

Marihuana Tax Act of 1937, Additional Statement of H. J. Anslinger
<http://www.druglibrary.org/schaffer/hemp/taxact/t10a.htm>

The Forbidden Fruit and the Tree of Knowledge: An Inquiry into the Legal History of American Marijuana Prohibition by Richard J. Bonnie and Charles H. Whitebread, II
<http://www.druglibrary.org/schaffer/library/studies/vlr/vlr3.htm>

Marihuana Conference, December 5, 1938
<http://www.druglibrary.org/schaffer/hemp/taxact/1938_mhc.htm>

Marihuana Conference, December 5, 1938, Part 2
<http://www.druglibrary.org/schaffer/hemp/taxact/mhc2.htm>

Marijuana Conference Index, December 5, 1938
<http://www.druglibrary.org/schaffer/hemp/taxact/mhc1.htm>

Hemp for Victory—U.S. Department of Agriculture, 1942
<http://www.druglibrary.org/schaffer/hemp/hemp4v.htm>

"War Demands More Hemp," June 1942
<http://www.druglibrary.org/schaffer/history/e1940/wardemands.htm>

The La Guardia Committee Report: The Marijuana Problem in the City of New York, 1944
<http://www.drugtext.org/library/reports/lag/lagmenu.htm>

Kefauver Hearings, 1951
<http://history.acusd.edu/gen/filmnotes/kefauver.html>

"Narcotic Menace Held Nation-Wide," June 18, 1951, *The New York Times*
<http://www.druglibrary.org/schaffer/history/e1950/narcmenace751.htm>

Legal References on Drug Policy—Federal Court Decisions on Drugs by Decade, 1930
<http://www.druglibrary.org/schaffer/legal/legal1930.htm>

Legal References on Drug Policy—Federal Court Decisions on Drugs by Decade, 1940
<http://www.druglibrary.org/schaffer/legal/legal1940.htm>

Legal References on Drug Policy—Federal Court Decisions on Drugs by Decade, 1950
<http://www.druglibrary.org/schaffer/legal/legal1950.htm>

NOTES

1. Biographical information about Anslinger is from the Federal Bureau of Investigation file on Harry J. Anslinger, U.S. Department of Justice, Washington, DC, hereafter referred to as "FBI file."

2. Richard Hofstadter, *The Age of Reform* (New York: Vintage Books, 1955), p. 176.

3. Harry J. Anslinger and Will Oursler, *The Murderers: The Story of the Narcotics Gangs* (New York: Farrar, Straus and Cudahy, 1961), p. 8.

4. Ibid., p. 8.

5. Anslinger's FBI file shows that early in his adult life, he had some connection with the Health School for Chronical Diseases in Harrisburg, Pennsylvania. I have been unable to determine whether this was as a student, an employee, or in some other capacity. This may be information worth uncovering, especially if it bears on Anslinger's attitude toward medicine and science in general. However, I am assuming nothing from this connection.

6. Special Inquiry, February 25, 1953, FBI file. The co-worker's name and the names of most other interviewees are blacked out of the file.

7. Anslinger and Oursler, *The Murderers,* p. 16.

8. Ibid., p. 20.

9. Ibid., pp. 19-20.

10. Ibid., p. 20.

11. Hearings on HR 11143 Before the Committee on Ways and Means, House of Representatives, *An Act to Create in the Treasury Department a Bureau of Narcotics,* 71st Congress, 2nd Session, March 7-8, 1930.

12. On July 1, 1930, Anslinger received two letters from Secretary of the Treasury Andrew Mellon: one, referring to Section 8 of the Prohibition Reorganization Act of 1930, naming him Assistant Commissioner of Industrial Alcohol, the other, referring to Section 2 of the "Act to create in the Treasury Department a Bureau of Narcotics," naming him acting Commissioner of Narcotics effective July 1, 1930, until further order. Anslinger Papers, Box 3, File 10. In 1961, Harry Anslinger donated his papers to Pattee Library at The Pennsylvania State University, University Park, Pennsylvania. They are housed in the Pennsylvania Historical Collections and Labor Archives. Further references to the collection will appear as AP.

13. Colonel Garland Williams, interview with author, February 17, 1990.

14. Anslinger and Oursler, *The Murderers,* p. 20. Demonstrating again that nothing is new, in one of his most publicized addresses, on September 5, 1989, President George Bush outlined his plan for dealing with what he called "the toughest domestic challenge we've faced in decades"—illicit drugs. Probably the most memorable moment of the president's address occurred when he held up a bag containing white powder and said, "This is crack cocaine, seized a few days ago by Drug Enforcement Administration agents in a park just across the street from the White House." One of the reasons this moment became memorable was that a few days later the press revealed that the drug buy was a set-up so that the president could have an effective prop for his speech. President George Bush, "The National Drug Control Strategy," Washington, DC, September 5, 1990.

15. These were some of the uses for marijuana, the blossom of the female hemp plant. The hemp plant itself was harvested in the United States from the moment colonists arrived from England, and almost no colonial farms were without the many products made from hemp: fabric for clothing, towels, bed linens, tablecloths, and even lace. The New England shipbuilding industry relied heavily on hemp for rope and sails. However, the Civil War made trade among southern hemp farmers and New Englanders impossible, and by the time the war ended, hemp had been replaced by other materials: iron proved stronger for cables, and jute was cheaper than hemp to produce. Government-controlled hemp production resumed during World War II, and the U.S. government even produced a film, "Hemp for Victory," to encourage farmers to grow hemp. However, the plant met with disfavor when it was linked with marijuana, which was linked with corrupting the nation's youth. The government put such rigid controls on hemp production that farmers discontinued growing it.

16. David F. Musto, *The American Disease: Origins of Narcotic Control* (New York: Oxford University Press, 1987 [1973]), p. 216.

17. Musto, *American Disease,* p. 217.

18. Ibid., p. 213. Also, in 1970, senior agent George White, who began working for the bureau around 1933, wrote, "During Anslinger's tenure, any Agent who thought he could get by merely by making numerous marihuana cases was quickly disillusioned. . . . This did not mean the Bureau thought marihuana was an innocuous drug, but only that it was felt we should operate on a basis of first things first." George White to John Kaplan, June 12, 1970, George White Papers, Stanford University, CA.

19. Harry J. Anslinger, "Necessity for the Enactment of the Proposed Uniform Narcotic Drug Act," *The Clubwoman GFWC,* October 1933, 6. AP, Box 1, File 11.

20. Hearing Before the Subcommittee of the Committee on Appropriations, House of Representatives, *Treasury Department Appropriations Bill for 1937,* 74th Congress, 2nd Session, December 14, 1935, 444. Through the 1930s, Anslinger advocated individual state adoption of the Uniform State Narcotic Drug Act that provided for state enforcement of an antimarijuana law. His success was measurable by the yearly increase in the number of states adopting the Uniform Law. Anslinger's year-end reports to the Treasury Department contained those figures: In fiscal year 1934, Anslinger reported that eight states had adopted the Uniform Law; fiscal year 1935, twenty-six states; fiscal year 1936, twenty-nine states; fiscal year 1937, thirty-eight states; and in fiscal year 1938, forty-one of the forty-eight states had adopted the Uniform Law. In addition, two of the remaining seven states and the territory of Hawaii had adopted "adequate narcotic legislation" of their own. Department of Treasury, *Annual Report of the Commissioner of Narcotics for the Fiscal Year Ended June 30, 1936* (Washington, DC: GPO), p. 169. Department of Treasury, *Annual Report of the Commissioner of Narcotics for the Fiscal Year Ended June 30, 1937* (Washington, DC: GPO), p. 176. Hearing Before the Subcommittee of the Committee on Appropriations, House of Representatives, *Treasury Department Appropriation Bill for 1941,* 76th Congress, 3rd Session, December 14, 1939, p. 433.

21. *Treasury Department Appropriations Bill for 1937,* pp. 446-447.

22. Ibid., p. 447.

23. F. W. Russe to H. J. Anslinger, January 30, 1936. AP, Box 3, File 4.

24. Representative Robert L. Doughton introducing HR 10586, *Congressional Record,* 74th Congress, 2nd Session, January 24, 1936, 80: 1002.

25. Harry J. Anslinger, "The Need for Narcotics Education," National Broadcasting Network, February 24, 1936. AP, Box 1, File 7.

26. Harry J. Anslinger, "The Narcotic Problem," The Attorney General's Conference on Crime, Washington, DC, December 13, 1934. AP, Box 1, File 7.

27. Anslinger's Hashishan/Assassin references, which he used during his entire appointment as commissioner, are too numerous to list in their entirety; some citations include the following: Harry J. Anslinger, "Marihuana," Columbia Broadcasting Network, October 23, 1937, AP, Box 1, File 7; House Committee on Ways and Means, *Taxation of Marihuana,* 75th Congress, 1st Session, April 27, 1937; Harry J. Anslinger, "Marihuana—A New Menace to Youth," *New York Herald Tribune,* Forum, October 25, 1938, AP, Box 1, File 7; Harry J. Anslinger, "American Leadership in Suppressing the Abuse of Dangerous Drugs," *The Nebraska Police Officer,* May 1939, 4, AP, Box 1, File 11; Anslinger and Oursler, *The Murderers,* p. 37.

28. *Oxford English Dictionary,* Second Edition, s.v. "assassin." See also *Enciclopedia Universal Ilustrada,* s.v. "asesinos."

29. Richmond P. Hobson, president of the World Narcotic Defense Association (WNDA) to H. J. Anslinger, February 3, 1936. Richmond P. Hobson, copy of letter to all members of the board of directors of the World Narcotic Defense Association, Inc., February 3, 1936. Resolution of the Board of Directors of the WNDA, Protesting Against the Submergence of the Bureau of Narcotics into a Sub-Division of the Division of the Secret Service. Richmond P. Hobson to Robert L. Doughton, February 3, 1936. Rowland Jones Jr., Washington representative of the National Association of Retail Druggists, to NARD members, February 27, 1936. Charles H. Tuttle, Executive Committee Chairperson of the WNDA, Address over WMCA in Observance of Narcotics Education Week, New York City, February 27, 1936, AP, Box 3, File 4.

30. *National Association of Retail Druggists Journal,* October 1, 1936, 405. AP, Box 3, File 4.

31. Harry J. Anslinger to H. E. Gaston, August 18, 1936. AP, Box 3, File 4.

32. Conference on *Cannabis sativa* L., Treasury Building, Washington, DC, January 14, 1937, p. 34. AP, Box 9, File 22.

33. Conference on *Cannabis sativa,* p. 35.

34. Harry J. Anslinger and Courtney Ryley Cooper, "Marijuana: Assassin of Youth," *The American Magazine,* July 1937, p. 150. AP, Box 1, File 11.

35. A. Parry, "Menace of Marihuana," *American Mercury* 36 (1935): 487-490; "Marihuana Menaces Youth," *Scientific American* 154 (1936): 150; William Wolf, "Uncle Sam Fights a New Drug Menace—Marihuana," *Popular Science Monthly* (May 1936); 14-15, 119-120; "Facts and Fancies About Marihuana," *Literary Digest* 122 (1936): 7-8; Wayne Gard, "Youth Gone Loco," *Christian Century* 55 (1938): 812-813; Henry G. Leach, "One More Peril for Youth," *Forum* 101 (1939): 1-2; "Danger," *Survey Graphic* 27 (1938): 221; F. T. Merrill, "Marihuana: Increasing Use and Terrifying Effects," *Journal of Home Economics* 30 (1938): 477-479; "Marihuana—A New Menace to U.S.," *The Nebraska Police Officer* (December 1938-January 1939). The last article is also in AP, Box 1, File 11.

36. Anslinger and Cooper, p. 150. AP, Box 1, File 11.

37. House Committee on Ways and Means, *Taxation of Marihuana,* 75th Congress, 1st Session, May 4, 1937, pp. 117-118.

38. Ibid., pp. 29-30.

39. AP, Box 3, File 3.

40. Harry J. Anslinger, "Peddling of Narcotic Drugs," March 20, 1933. AP, Box 8, File 4.

41. *Marihuana Tax Act: Law and Regulations Relating to the Importation, Manufacture, Production, Compounding, Sale, Dealing In, Dispensing, Prescribing, Administering, and Giving Away of Marihuana,* 75th Congress, 1st Session, 1937.

42. Hearings Before the Subcommittee of the Committee on Appropriations, House of Representatives, *Treasury Department Appropriation Bill for 1944,* 87th Congress, 1st Session, December 10, 1942, p. 1.

43. The FBN's fiscal year 1943 budget was $1,289,060, and the fiscal year 1944 estimate was $1,200,000. *Treasury Department Appropriation Bill for 1944,* p. 476.

44. Ibid., p. 487.

45. H. J. Anslinger, Memo, April 9, 1941. AP, Box 2, File 21.

46. *Treasury Department Appropriation Bill for 1945,* p. 537.

47. Ibid., pp. 532-533.

48. Eric F. Goldman, *The Crucial Decade—And After: America, 1945-1960,* Revised Edition (New York: Random House, 1960), pp. 78-79, 113.

49. Ibid., pp. 132-133.

50. Hearings Before the Subcommittee of the Committee on Appropriations, House of Representatives, *Treasury Department–Post Office Appropriations for 1952,* 82nd Congress, 1st Session, February 16, 1951, p. 272.

51. Representative Jacob K. Javits, "Marihuana Makes Youthful Criminals," February 27, 1950, *Appendix to the Congressional Record,* 81st Congress, 2nd Session, 96:A1521.

52. Ibid.

53. "Youth Narcotic Use Growing," *The New York Times,* November 5, 1950, p. 44.

54. "Narcotics Users Warned," *The New York Times,* February 2, 1951, p. 21.

55. "Narcotics Unit Cites Lure to Teen-Agers," *The New York Times,* March 2, 1951, p. 19.

56. Anslinger had previously testified before the Kefauver Committee on June 28, 1950. A number of FBN agents also testified as the committee held hearings in various cities across the country. See Estes Kefauver, *Crime in America* (Garden City, NY: Doubleday & Co., Inc., 1951), p. 20.

57. Senate Special Committee to Investigate Organized Crime in Interstate Commerce, testimony of Harry J. Anslinger, March 27, 1951. Anslinger's statement is housed in the Anslinger files in the Drug Enforcement Administration Library, Washington, DC.

58. Ibid.

59. Ibid.

60. Kefauver, *Crime in America,* p. 20.

61. Ibid., pp. 326-327.

62. Ibid., pp. 325-326.

63. Hearings Before a Subcommittee of the Committee on Ways and Means, House of Representatives, *Control of Narcotics, Marihuana, and Barbiturates,* 82nd Congress, 1st Session, April 7, 14, 17, 1951, pp. 208.

64. Ibid., p. 209.

65. "Narcotic Menace Held Nation-Wide but Gravest Here," *The New York Times,* June 18, 1951, p. 1.

66. Ever since 1942, when New York Mayor Fiorello H. LaGuardia, who supported Anslinger for the commissioner's position in 1930, authorized publication of *The Marijuana Problem in the City of New York,* a scientific study that contradicted Anslinger's public marijuana claims, Anslinger frequently pointed to New York City as the hotbed of narcotics traffic and use. See Rebecca Carroll, "The Rhetoric of Harry J. Anslinger, Commissioner of the Federal Bureau of Narcotics, 1930 to 1962" (doctoral dissertation, University of Pittsburgh, 1991), Chapter 3.

67. "Narcotic Menace Held Nation-Wide but Gravest Here," p. 1.

68. "Teen-Age Dope Addicts: New Problem?" *U.S. News and World Report,* June 29, 1951, p. 18. Oregon Representative Homer Angell submitted this article as part of his remarks in support of HR 4593, a bill that he introduced that would "impose life sentences on certain major criminal addicts with death sentences in certain cases where these diabolical criminals are convicted of peddling these habit-forming drugs to persons under the age of 21 years and thereby making them slaves of the drug habit." See *Congressional Record,* 82nd Congress, 1st Session, June 26, 1951, 97:7148-7150.

69. "Teen-Age Dope Addicts: New Problem?" p. 18.

70. Ibid., p. 19.

71. "Youth: High and Light," *Time,* February 26, 1951, pp. 24-25.

72. "Drug Peddling, the Dirtiest Crime," *Pathfinder,* June 27, 1951. This article became part of the June 26, 1951, *Congressional Record,* 82:7150.

73. "Child 'Dope' Addicts," *Science News Letter,* July 14, 1951, p. 29.

74. Harry J. Anslinger, "Control of the Traffic in Narcotic Drugs," *The Merck Report,* October 1951, pp. 33-35.

75. Ibid., p. 35.

76. An examination of *The Gallup Poll* from 1930 to 1990 reveals that in response to the questions, "What do you think is the most important problem facing this country today?"; "What do you think is the most important problem facing the country in the next year?"; "What is the most important problem you and your family face today?"; "What is the most important issue that Congress should take up?"; and "What is the most important domestic problem?" no reference to drug use appeared in Gallup's lists ranging from three to fourteen items until July 14, 1972, at which time it appeared third at 9 percent, following Vietnam at 25 percent and economy/cost of living at 23 percent. At no time during Anslinger's tenure did drug use appear on any list of important problems in the Gallup Polls.

77. "Narcotics Penalty Stiffened in House," *The New York Times,* July 17, 1951, p. 16. Copyright © 1951 by The New York Times Co. Reprinted with permission.

78. Ibid., p. 16. Copyright © 1951 by The New York Times Co. Reprinted with permission.

79. According to Washington attorney Rufus King, Representative Hale Boggs "was Anslinger's biggest puppet up there [on Capitol Hill] on the House side." King said that Boggs "made something of a career of being tough on dope fiends." Rufus King, interview with the author, January 11, 1990. In his book *The Drug Hang-up: America's Fifty-Year Folly* (Springfield, IL: Charles C Thomas, 1972), pp. 270-273, King writes that Boggs continued to present pieces of legislation with increasing

punitive measures. Later, Anslinger began to become impatient with Boggs when the latter kept insisting that the FBN police barbiturates, an area in which Anslinger did not wish to involve his bureau.

80. *Congressional Record,* October 20, 1951, 82nd Congress, 1st Session, 97: 13675.

81. Ibid., p. 13676.

82. "Truman Signs Bill for Narcotics War: Stiff Prison Terms Mandatory in Repeat Offenses," *The New York Times,* November 3, 1951, p. 33. Copyright © 1951 by The New York Times Co. Reprinted with permission.

83. John M. Conly, "'Shoot First' Is Nationwide Slogan for Raids on Dope Peddlers," *Pathfinder,* January 23, 1952, p. 25.

84. *Historical Statistics of the United States, Colonial Times to 1970* (Washington, DC: U.S. Department of Commerce, Bureau of Census; 1975), Volumes I and II.

85. David F. Musto, Yale psychiatrist and noted historian of American drug history, pointed out that narcotics use reached an all-time national low during World War II but that use was never completely eradicated. The use of imported drugs such as opium and heroin declined because of wartime difficulties in international traffic. However, because marijuana was a domestic drug, it could have posed a particular problem for the FBN during the war years. But, Musto explains, from 1937 until 1960, the American attitude toward drug abuse was one of "extreme intolerance." David F. Musto, "Are There Lessons from America's Drug History?" Lecture at University of Pittsburgh, Pittsburgh, Pennsylvania, November 3, 1988. In his book, Musto wrote of that period, "Once commonplace, narcotics use waned into little more than rumors about this physician or that movie star. Personal knowledge of a 'dope fiend' was unusual for the vast majority of Americans by the 1950s." Further, Musto commented on the mind-set of many Americans toward drugs: "The consensus on drugs—intolerance toward the use or advocacy of narcotics—was well-established by the mid- or late 1920s. As a result, members of the generation born in the 1920s grew to maturity with diminishing direct knowledge of, but a great deal of animosity toward, the substances." *American Disease,* p. 251.

86. Figures for yearly total number of convicts with incarceration sentences for drug offenses are from *Historical Statistics of the United States, Colonial Times to 1970* (Washington, DC: U.S. Department of Commerce, Bureau of the Census, 1986), Volumes I and II.

87. "Truman Signs Bill for Narcotics War: Stiff Prison Terms Mandatory in Repeat Offenses." Copyright © 1951 by The New York Times Co. Reprinted with permission.

Chapter 4

The Narcotic Control Act Triggers the Great Nondebate: Treatment Loses to Punishment

Rebecca Carroll

In the early 1950s, law enforcement officials reported a sharp increase in juvenile delinquency. Legislators and law enforcement officials explained this alarming trend by connecting juvenile delinquency with narcotics use. Anslinger agreed, "This epidemic of narcotic addiction among younger people is primarily an extension of a wide-spread surge of juvenile delinquency."[1] Yet Anslinger maintained that, overall, teenage drug use was not only very much under control, but was actually being exaggerated by groups of alarmists. In the December 1952 issue of *International Criminal Police Review,* Anslinger wrote,

> In some cases that [public concern] has amounted to a near hysteria. Always when the public becomes extremely interested in a subject there are persons who take extreme views. The narcotic situation in this country is serious. It should not be minimized but we are by no means in a helpless situation. . . . Acute public concern always brings its problems.[2]

Apparently not satisfied that first-time narcotics convictions carried no mandatory minimum sentence and that first-time offenders could be eligible for parole or probation, Anslinger again focused his efforts on peddlers selling to young people. A 1953 headline from the *Altoona Mirror* read: "Anslinger Calls for Stiffer Penalties on Dope Peddlers to Protect Youth of Nation."[3] The same day, November 23, 1953, *The Washington Post* ran this headline: "Anslinger Asks Senate

Action on Addict Bill; Hospitalization Law Still Needed, He Says; Drug Use Declines."[4] Both articles covered Anslinger's testimony before the Senate Juvenile Delinquency Subcommittee on November 23, 1953. The *Mirror* reported that in his testimony, Anslinger said that drug addiction among teen-agers appears to be declining but is still far too widespread for complacency. He called for stiffer state and federal penalties against drug peddlers to help break up the rings of racketeers preying on the nation's youth.[5]

By the mid-1950s, sociologist Alfred R. Lindesmith would offer a different explanation for the abuse of narcotic drugs by young people. Lindesmith cited "Dr. E. Bishop, a noted authority on drug addiction," who as early as 1921 "commented on the trend toward juvenile addiction and ascribed it to the 'prohibition' control technique." Lindesmith wrote,

> The disappearance of the clinics marked the final triumph of the "prohibition" idea and the complete removal of the control issue from the medical domain. The drug problem is what it is today as the result of these moves by the government.[6]

For young people, the result was an increase in addiction. Lindesmith cited the Federal Bureau of Investigation's *Uniform Crime Report:* "In 1932, for instance, only fifteen per cent of narcotic law violators were under twenty-five years of age; in 1940, the figure had reached twenty-six percent; today [1956] is a little under fifty per cent."[7]

Early in 1955, the U.S. Senate Subcommittee on Illicit Narcotics Traffic was formed.[8] On May 31, 1955, Subcommittee Chairman Senator Price Daniel reported to Congress that hearings would begin in Washington on June 2, 1955, and that Commissioner Anslinger had agreed to "pinpoint by city and State the actual figures showing the number of drug addicts and narcotic law violations."[9]

On the first day of the hearings, Anslinger testified before the subcommittee, whose members questioned him in several areas: effects and availability of various drugs, treatment programs, the issue of legalization, and numbers of addicts. In reference to marijuana, Anslinger stated, "in recent years marihuana smoking has become an increasing problem."[10] Further, he distinguished between marijuana, a habit-forming drug, and other addiction-forming drugs, saying, "You can break the marihuana habit probably in a day." Seeking clarification of statements that Anslinger made in his book, *The Traffic in Narcotics,* Senator Daniel questioned the economic impact that even a limited marijuana user had on a community, asking, "Marihuana can cause a person to commit crimes and do many heinous things; is that not correct?" And the commissioner responded, "That is correct. It is a dangerous drug, and is so regarded all over the world," adding that the marijuana problem had become a worldwide concern and that the United Nations was in the process of studying it. However, Anslinger did not mention his own membership on the United Nations Narcotics Commission. Anslinger added that while the United States still had a legitimate hemp business, the Department of Agriculture was still "attempting to find a plant which does not have the resinous or the poisonous properties."[11]

Later, when the testimony turned to treatment of addicts, Anslinger staunchly maintained his position that compulsory hospitalization was the best course of action. Exploring another form of treatment—legalized drugs through clinics—Idaho Senator Herman Welker referred to the "bleeding hearts" who advocated such an approach. Welker said that those people were "acting through sympathy of the poor addict," yet this addict might also have been a peddler who "by his sale can liquidate and ruin thousands of people in the community." Anslinger replied that not only could the committee expect propaganda from "the drug addict to legalize his supply," but also from "a considerable minority group . . . [that] advocates [ambulatory treatment], including some attorneys [referring to Rufus King, who would testify later] who think that is the answer to the problem."[12] Finding fault with this minority—which happened to be a joint committee of the American Bar Association and the American Medical Association—Anslinger used an analogy, saying that while this group supported legalized free treatment for addicts, it would not set up bars for alcoholics. He said, "Certainly I just could not conceive of the Government setting up legalized stores to sell marihuana cigarettes or cocaine."[13]

In a tactical move, Anslinger recommended "that the committee hear some proponents and hear some opponents." Senator Daniel agreed because "so many outstanding people [are] advocating the free treatment of addicts and the free dosages of drugs."[14] Later, however, the New York *Daily News* would chastise Anslinger for how little regard, in fact little tolerance, he demonstrated for the opposition.[15]

At this point, Anslinger asked if he might introduce a visitor, Robert Curran, legal adviser to the Department of National Health and Welfare in Canada. No ordinary visitor, Curran happened to hold dear the same beliefs as Anslinger in regard to ambulatory treatment. Apologizing for not having a formal written statement because he was not aware that he would be asked to speak, Curran took the floor, expressing his pride in sharing the same platform with the U.S. commissioner, whom Curran regarded as "an outstanding authority in this field."[16]

At this hearing, Anslinger offered for the first time an explanation of the 60,000 U.S. addicts (mostly using heroin), a number that he cited many times through the 1950s. He said that "based upon the actual names and records reported to the Bureau by local, State, and Federal authorities since January 1, 1953," the count had yielded approximately 1,000 addicts each month, "with total of 28,514 counted for the 28 months ending April 30, 1955." Anslinger claimed that *every* addict came to the attention of the FBN within approximately two years because "he is engaged in criminal activities of one form or the other."[17] Therefore, "we are fairly certain that this count that we are getting now will go up to about 60,000 probably within a 5-year period." He substantiated this figure by pointing out that FBN numbers were comparable to "the figures for rejections of drug addiction by the Selective Service." For certain age groups, Anslinger said, Selective Service rejections could be plotted closely to the FBN's addict figures, showing that "you can be reasonably certain that this count is going to be correct, and it should not go much over 60,000."[18]

However, this contrasted sharply with evidence in Alden Stevens' November 1952 *Harper's Magazine* article in which he cited the New York City Mayor's Committee on Narcotics. According to the committee, the number of addicts in New York City alone during the summer of 1951 was between 45,000 and 90,000. Stevens stated that this was an estimate and that "the FBN . . . says, and all experts agree, that there are no reliable figures on the total number of addicts."[19]

Yet for many years, Anslinger repeated his claim that the United States housed 60,000 addicts. As early as December 14, 1939, Anslinger claimed that the number of drug addicts in this country declined from one in every 1,000 persons in 1924 to one in every 3,000 persons in 1939.[20] Perhaps thinking that a larger *total* number of addicts would carry more authority, Anslinger changed his approach in the 1950s to claim that 60,000 drug addicts resided in the United States, and he offered an objective method for arriving at that figure: based on a reporting system that the FBN initiated with local and state enforcement agencies, the names of 1,000 addicts each month reached the FBN, or 12,000 each year, which, as he estimated, would result in 60,000 in five years. While almost impossible to prove, Anslinger probably established that figure by looking at the U.S. population at that time—165,275,000 in 1955—and dividing that number by 3,000, because his earlier claim was that one person in 3,000 was addicted. The quotient was 55,092, and by simply rounding that figure up to the nearest 10,000, he conveniently arrived at 60,000. Sociologist Lindesmith would later describe Anslinger's figure of 60,000 addicts as "a guess; its main virtue is that it is the lowest offered."[21] Despite whatever disputes occurred over the numbers of U.S. addicts, Anslinger's statistics seemed credible because he had over twenty years' experience as the narcotics commissioner. Of all the debate participants, Anslinger was in the position most likely to have accurate figures.

Generally when Anslinger spoke before any legislative or investigative body, the print media covered his testimony, and this event was no exception. From his substantial amount of testimony, *The New York Times* chose to headline Anslinger's estimated number of addicts. The June 3, 1955, issue carried an article titled "Narcotics Users Are Put at 60,000," with a subheadline showing that 9,458 of the addicts were located in New York.[22] Quoting Anslinger, the *Times* noted that the nation's failure to curb drug addiction lay primarily with "thc legislators and other officials" who had been lax in creating and enforcing stringent narcotics laws. As usual, Anslinger's twofold approach to reducing the number of drug abusers involved heavier sentences and compulsory hospitalization.

Shortly after Anslinger's testimony became public, congresspersons began to react by introducing numerous bills for narcotics legislation with even greater penalties. On June 9, 1955, Representative William E. McVey of Illinois introduced House Joint Resolution 188, which provided for a five-year minimum sentence for first offenders selling narcotics, a ten-year minimum sentence for second offenders, and life imprisonment for third offenders. McVey stated, "These penalties are more severe than anything on the statute books at present and should be a deterrent to those who try to make profits from these deadly drugs."[23] On June 20, 1955, Republican Representative Henry J. Latham of New York introduced House Bill 6921, which provided for the death penalty for anyone convicted of selling narcotics to a minor.[24] On July 27, 1955, Democratic Representative Jere Cooper of Tennessee introduced House Resolution 7018 to enable the secretary of the Treasury Department, which housed the FBN, to authorize subpoenas for evidence and witnesses in narcotics cases.[25]

In July 1955, when the Senate subcommittee's investigation centered on the narcotics problem in Washington, DC, Democratic Senator John J. Sparkman of Alabama addressed the Congress, saying, "The District of Columbia alone needs at least a dozen new laws, or amendments to old laws, to cope with the narcotics problem here in the Nation's Capital."[26] Sparkman did not elaborate on what exactly the dozen-plus laws should address. These bills were either a testament to Anslinger's ability to persuade numerous lawmakers of the value of punishment or evidence that congresspersons would latch onto any issue that would please their increasingly frightened constituency, ensuring their chances of reelection.

BEYOND BOGGS: THE NARCOTIC CONTROL ACT OF 1956

On September 18, 1955, *The New York Times* ran an article titled "Stiffer Law Asked by Narcotics Chief," in which Anslinger referred to the Boggs Act, saying that "the four-year-old law providing stiff penalties for narcotic peddling had helped to cut such traffic 'way, way down' in some areas." Even though the average prison sentence had increased from eighteen months to four years since 1949, Anslinger advocated a mandatory sentence for first offenders because "veteran 'pushers' who are subject to a minimum of five years on a second offense and ten years on a third, 'are having the peddling done by the fellow who hasn't had a first conviction.'" Anslinger then praised the new Ohio state law that provided for a minimum sentence of two years "for inducing or trying to induce a person to use or administer narcotics unlawfully and a minimum of thirty years for inducing a minor to use narcotics."[27]

Lindesmith, too, commented on the skill of the veteran pusher to avoid getting caught. However, he placed the blame on the government:

> The huge illicit traffic, directed for profit by non-addicted lords of the underworld, has become the focal point of new infection. These men are rarely apprehended or punished; it is the user, exploited by the system, who suffers the major portion of the heavy penalties that are imposed. Police suppression, by increasing the danger of distribution and reducing supplies, keeps up prices and profits.[28]

A September 17, 1955, *New York Times* article, "Free Narcotics to Be Pondered," announced the upcoming visit of Senator Daniel's subcommittee to New York to discuss the New York Academy of Medicine's proposal "that addicts be given free narcotics at state clinics."[29] One witness at the investigation who attempted to quash the ambulatory treatment idea was Commissioner Anslinger, who called the proposal "propaganda that the Government sell poison at reduced prices to its citizens."[30] Anslinger attacked the plan and its proponents, claiming unequivocally and prematurely that the plan would fail.

Proponents for the plan argued that some addicts were, in fact, incurable, and an ambulatory approach would benefit them in two ways. First, it would enable them to maintain as "normal" an existence as possible, perhaps even holding a job, something that would be impossible with compulsory hospitalization. Second, the ambulatory approach would keep such users out of the hands of the narcotics peddlers and thereby eliminate part of the market for illegal drugs.

Two New York Medical Society physicians, Dr. Andrew A. Eggston and Dr. Herbert Berger, presented the society's plan for the national ambulatory clinic program. The society's formal statement made clear that the society hoped to work closely with the FBN in administering the program, which involved nine well-delineated recommendations providing maximum safeguards to protect the addicts, the physicians, and the clinics. The society's plan was formulated on the belief that "addicts are patients with mental disturbances, and should be treated by physicians, presenting a medical problem with very often the legal and criminal problem appearing secondarily."[31]

Also present were two representatives from the American Medical Association, who said that, while the AMA had not yet changed its opinion of many years opposing ambulatory treatment, "it was about to begin a reappraisal of the whole narcotics problem in cooperation with the American Bar Association."[32] Neither the society's recommendation nor the AMA's announcement was good news for Anslinger, who vehemently opposed any course of action other than punitive measures and compulsory hospitalization, both of which the New York Medical Society opposed.

To rebut, Anslinger used pathos-filled quotations from unnamed addicts and convicts who claimed that the longer jail terms prohibited them from participating in the drug trade. He quoted unnamed parent-teacher groups and women's groups who espoused beliefs identical to his own. He also continued to quote as a third, unbiased party the United Nations Narcotics Commission, disregarding the fact that he was an active, respected, and vocal member of the commission. Thus, when Anslinger quoted the United Nations Narcotics Commission, he was essentially quoting himself.

Commissioner Anslinger was probably even less delighted to see the next witness, attorney Rufus King, representing the American Bar Association (ABA). King began by submitting a recommendation unanimously approved by the ABA House of Delegates, calling for a thorough reexamination of the provisions of the Harrison Narcotics Act of 1914.

In addition, King testified that the House of Delegates authorized the ABA to initiate meetings with the AMA to form a joint committee that would study "the entire narcotic drug traffic and the related problems it encompasses."[33] King reported that the AMA welcomed the invitation from the ABA. The joint committee would "back off and take a long look at some of these questions which are being raised here, and back up their study, where necessary, with research."[34] None of these representatives made claims at this point; all were guarded and speculative about their plans, looking to the future to supply the ultimate answers to the problem.

Despite the testimony of four physicians and King, the Daniel Subcommittee concluded that "the academy plan was unworkable and that such clinics would only tend to make the addiction problem worse." In addition, *The New York Times* quoted Daniel as saying that he was "convinced we are never going to lick the problem of narcotics until we take the drug addicts off the street," supporting Anslinger's position on compulsory hospitalization. Daniel recommended that "addicts discharged from hospitals as cured be put on parole so they could be sent back for compulsory treatment if they reverted to drug use." The ambulatory treatment method apparently received no real consideration, despite the testimony of these expert witnesses.[35]

On October 24, 1955, Representative Hale Boggs returned to the discussion when Senator Daniel announced that his subcommittee would advocate "the death penalty for narcotics smugglers and some narcotics peddlers."[36] However, Boggs opposed such a measure, and as *The New York Times* reported, Boggs had "initial Congressional responsibility in devising any changes in the narcotics laws" because the narcotics laws are part of the Internal Revenue Code "and all revenue bills must originate in the House." Instead of the death penalty, Boggs reported he intended to sponsor a bill that prohibited parole or suspended sentences for first-time narcotics law offenders, as was currently allowed. In addition, the bill would provide for a mandatory five-year sentence for first offenses and a mandatory ten-year sentence for repeaters. With greater mandatory sentences, Boggs said, "I am absolutely convinced that we can dry up the illicit traffic in dope."[37]

The Daniel Senate Subcommittee on Illicit Narcotics Traffic surprised everyone by concluding its proceedings about a month ahead of schedule, an event *The New York Times* described as "somewhat unusual in the run of Congressional investigations."[38] That announcement was made on January 8, 1956. Probably not coincidentally, Daniel announced on April 9, 1956, his intention to run for governor of Texas.[39] The efficiency of the Daniel subcommittee and the following two additional factors put gubernatorial candidate Daniel in a favorable light. *Time* magazine reported that rather than the usual requests for more time and more funding, the subcommittee "completed its task before the January 31 deadline" and would "return to the Senate's investigatory fund more than $15,000 of the $50,000 allotted to it last spring." Finally, the subcommittee would "present precise legislation, rather than mere recommendation, before the deadline." In total, Daniel's subcommittee performed efficiently and assertively, two desirable gubernatorial qualities from a candidate promising "to give the people 'simple honesty and moral integrity in government.'"[40] Anslinger would share the benefits.

After thirty-seven days of open hearings in Washington, Philadelphia, New York, Austin, Fort Worth, Dallas, Los Angeles, San Francisco, Chicago, Detroit, and Cleveland, which resulted in 8,667 pages of testimony from 345 witnesses, the subcommittee presented ten findings and made thirteen recommendations to Congress on January 9, 1956.

No doubt Anslinger was pleased with the findings that supported his long-held beliefs. Among these findings were the following points. First, the number of drug addicts in the United States had tripled since World War II. Second, drug addiction was responsible for approximately 50 percent of crime in the United States. Third, the primary sources of drugs coming into the country were the People's Republic of China, Turkey, Lebanon, and Mexico, and China had purposefully set out to demoralize free nations by infiltrating them with drugs. Fourth, 90 percent of the overland heroin and marijuana coming into the United States was from Mexico. Finally, criminal laws and procedures were insufficient, and penalties for narcotic violations were far too weak.

Surprisingly, none of the findings dealt specifically with teenage addicts. When asked about the problem of teenage drug abuse, Senator Daniel responded, "the nationwide situation with reference to juveniles is encouraging." However, he added that even though teenage addicts comprised only 13 percent of the total, this was still too high. Likewise, the subcommittee's recommendations would please the commissioner: five recommendations dealt with initiating international treaties for controlling narcotics production and traffic; five recommended changes in narcotics laws, including increased penalties, more liberal search and seizure provisions, and tighter constraints on the mobility of known narcotic addicts; another recommended increased funding for enforcement agencies such as the FBN and specifically suggested the addition of fifty FBN agents. With regard to the final recommendation, Senator Daniel praised "the able direction of Commissioner Harry J. Anslinger, [who] has done a splendid job in holding the narcotics traffic to its present level, considering its limited personnel and operating funds."[41]

On April 30, 1956, Senator Daniel submitted Senate Bill 3760 to the Senate Judiciary Committee with provisions similar to the subcommittee's recommendations in January.[42] On May 15, 1956, the Senate Judiciary Committee unanimously approved the bill.[43]

When the Narcotic Control Act of 1956 (HR 11619) went before Congress on June 19, 1956, it contained a recommendation on narcotics education that Anslinger could have written. It stated that the subcommittee was "convinced that the public generally does not fully understand the viciousness of drug addiction nor the seriousness of the proportions of this addiction." Further, it stated that the subcommittee gave serious consideration to instituting a narcotics education program in the schools; however, additional

> careful consideration . . . has led to the conclusion that it would tend to arouse undue curiosity on the part of the impressionable youth of our Nation. . . . Many young persons, once their curiosity is aroused, may ignore the warnings and experiment upon themselves with disastrous consequences.[44]

The subcommittee cited two authorities as references for this recommendation: the United Nations Commission on Narcotic Drugs, of which Anslinger was a member, and the narcotics commissioner himself. In other words, the subcommittee, perhaps unwittingly, may have cited only one source.

When President Eisenhower signed the Narcotic Control Act on July 18, 1956, he put into place the heaviest penalties for U.S. narcotics law violations to that point. The first conviction for possession of narcotics carried a prison sentence of two to twenty years with the possibility of parole or probation. The second conviction for possession and the first conviction for selling narcotics carried a mandatory prison sentence of five to twenty years with no parole or probation. And the third conviction for possession and the second or subsequent conviction for selling narcotics carried a mandatory prison sentence of ten to forty years with no parole or probation. In addition, the law provided for a $10,000 fine and a minimum sentence of ten years in prison for selling heroin to a person under eighteen, and the maximum penalty for selling heroin to a person under eighteen allowed a jury to recommend the death penalty.

Six months later, when Anslinger testified before the 1958 House Appropriations Subcommittee on February 4, 1957, he probably felt deserving of New Jersey Representative Gordon Canfield's opening remarks:

> Dr. Anslinger, because you are the No. 1 American general in this fight against narcotics addiction in our United States and because you are what I truly believe to be the world's greatest authority on narcotics, serving as you do as our representative in the United Nations, I am pleased beyond words to have your report today indicating chiefly that you believe that this year marks the turning point in our big fight against this awful menace. You say that because you are beginning to get from the Congress of the United States, supported by many States in our Union, the legislative weapons necessary to win this war.[45]

After twenty-six years, Anslinger was the recognized authority on narcotics. By controlling the discussion on narcotics, Anslinger controlled the policy on narcotics. The "war" would continue with a new enemy: anyone in favor of open discussion of narcotics policies.

LET THE DEBATE BEGIN?

In February 1955, the ABA House of Delegates passed a three-part resolution calling for, first, cooperation between the ABA and the AMA to study jointly "the narcotic drug traffic and related problems"; second, investigation through the American Bar Foundation for funding of the joint study, and third, "re-examination of the Harrison Act, its amendments and related enforcement and treatment policies and problems" by the Congress.[46]

The AMA agreed to work with the ABA in "truly a joint undertaking from the outset."[47] The six-member committee was formed in 1955 and remained intact through the process, working together on the *Interim Report,* which it presented to the two parent organizations in 1958, and on the *Final Report* presented in February and June 1959 and published in 1961. The three ABA members of the committee were Judge Edward J. Dimock, U.S. District Court; attorney Abe Fortas; and attorney Rufus King. The three AMA members of the committee were Dr. R. H. Felix of the National Institute of Mental Health; Dr. Isaac Starr of the University of Pennsylvania; and attorney C. Joseph Stetler, council for the AMA. In addition, at its meeting in December 1956 in Washington, DC, the joint committee members selected a director to oversee the research project. Judge Morris

Ploscowe agreed to serve. He had formerly directed the ABA's Commission on Organized Crime.[48]

In March 1957, the joint committee met in Philadelphia with Judge Ploscowe and agreed to keep three aims central while developing research projects: first, to examine existing sources and material for reliability; second, to design research projects that would "remedy deficiencies in present knowledge of the field"; and third, to attempt to draw conclusions by analyzing existing sources without doing further research.[49] The committee wanted to avoid duplicating previous research efforts to conserve its $15,000 grant from the Russell Sage Foundation.

The ABA-AMA Joint Committee moved intentionally slowly. The *Interim Report* showed that the joint committee "noted that a considerable amount of research might be required before final conclusions could safely be reached."[50] "But until more is learned about the narcotic problem—until at least some of the proposed research is completed—the Joint Committee believes it should proceed slowly."[51] The *Interim Report* ended neither by making conclusions nor by providing a timeline for forthcoming conclusions. The committee would continue "at its present pace until it is satisfied that the assignment given to it has been fully carried out."[52]

On the other side of the debate, defending the established definition of a drug user as a criminal was the FBN Advisory Committee, whom Anslinger described as "distinguished experts in the field of narcotics."[53] Chaired by Representative Hale Boggs, the FBN Advisory Committee consisted of twenty-six people, including narcotics agents; judges; pharmacologists; lawyers; psychiatrists; and Price Daniel, then governor of Texas. According to ABA-AMA Joint Committee member Rufus King, the FBN Advisory Committee "never actually convened."[54]

In the introduction to the *Interim Report* and *Final Report,* which were published together with appendixes in 1961 under the title *Drug Addiction: Crime or Disease?,* Lindesmith outlined the central problem: that those interested in the issue of drug addiction held two distinctly different opinions. First, the FBN and its supporters

> regard addiction to narcotic drugs as an activity that is properly subject to police control. . . . The advocates of this punitive approach argue that crimes committed by addicts are a direct

> result of the drug; they also contend that most addicts were criminals before they became addicted.[55]

The second opinion held that

> many addicts become criminals in order to get money to buy drugs. . . . From this point of view, drug addiction is primarily a problem for the physician rather than for the policeman, and it should not be necessary for anyone to violate the criminal law solely because he is addicted to drugs.

Further, to "remove the stigma of criminality from addiction . . . would aid materially in undermining the illicit traffic."[56]

Finally, Lindesmith listed "the two basic needs in such an enterprise . . . full and free investigation and full and free discussion."[57] As the events unfolded, these basic needs never were realized.

Interim Report

The six-member ABA-AMA Joint Committee completed its first draft of the *Interim Report* in February 1958, making five recommendations. Because the debate focused primarily on the first recommendation, only it will be discussed here: the joint committee, acknowledging the controversy surrounding the "so-called clinic approach to drug addiction," recommended "an outpatient experimental clinic for the treatment of drug addicts." The committee suggested the District of Columbia as the clinic site because its jurisdiction was "exclusively federal" and because it was "immediately accessible to both law-enforcement and public health officials."[58]

The ABA-AMA Joint Committee based these findings on the research presented in two extensive appendixes: Appendix A, "Some Basic Problems in Drug Addiction and Suggestions for Research" by joint committee director Morris Ploscowe, and Appendix B, "An Appraisal of International, British and Selected European Narcotic Drug Laws, Regulations and Policies," by committee member Rufus King.

Appendix A was a 120-page, twelve-part, thorough and well-documented analysis that included the history of the Harrison Narcotics Act of 1914; the World Health Organization's definition of drug addiction; the extent, nature, characteristics, and effects of drug addiction; the psychiatric, psychological, and social factors of drug addiction; the relationship between addiction and crime; methods of treatment; relapse and rehabilitation of drug addicts; the suggested clinics for legal narcotics distribution; and proposals for research. Much of the data in this document contradicted the FBN's public statements.[59]

Appendix B was a thirty-five-page analysis of international treaties, agreements, and narcotic drug policies of Great Britain, Denmark, Sweden, Norway, Belgium, and Italy. Consistent with arguments he made through the years, King reasoned that these countries had fewer addicts and less drug-related crime because of the following conditions: first, the medical profession policed itself mostly through Ministries of Health. Should a doctor be suspected of over-prescribing, the Ministry of Health would notify the doctor, and if it determined that a problem did exist, it would counsel and observe the doctor without the involvement of law enforcement agencies. Second, the doctors who administered to addicts were not criminals. Third, addicts were not viewed as criminals and were not handled by law enforcement but were treated by the medical profession as patients.

This document was the foundation for the conversation the joint committee wished to begin with interested parties. Almost all of the principles and policies contained in the joint committee's *Interim Report* strongly opposed Anslinger's established principles and policies that he had worked hard to protect through almost three decades. Obviously the commissioner would not respond favorably, and the exchange immediately became explosive.

Invitations for Discussion

On February 24, 1958, Morris Ploscowe sent Anslinger a copy of the *Interim Report* with appendixes, expressing his pleasure in presenting the material to him and also welcoming "comments and suggestions" from the commissioner. Further, Ploscowe asked that Anslinger choose three or four dates that would be convenient for him to meet with the ABA-AMA Joint Committee to discuss the *Interim Report.* Ploscowe concluded his brief letter by writing, "I look forward to hearing from you and also look forward to a discussion of narcotics problems with you."[60] The numerous letters of invitation that followed this one indicated that the ABA-AMA Joint Committee was eager to receive Anslinger's input on the document.

Although Anslinger began his March 4, 1958, response to Ploscowe by expressing gratitude for the materials, his second sentence and the remainder of his letter were antagonistic and divisive. Anslinger wrote, "I find it incredible that so many glaring inaccuracies, manifest inconsistencies, apparent ambiguities, important omissions, and even false statements could be found in one report on the narcotic problem." Thus, in one sentence, Anslinger simply dismissed the entire *Interim Report* without explanation. Later in his letter, he called the authors of the *Interim Report* "unquestionably prejudiced." Despite Anslinger's assurance that "we do not wish to censor the report" and his invitation to "sit down with our people to make necessary corrections," his actions and statements flatly contradicted those sentiments. Anslinger closed his letter by assuring Ploscowe that regardless of the policy—"enforcement, clinics, the British System, hospitalization, or penal provisions"—the ABA-AMA Joint Committee should "submit for consideration a factual document" so that those responsible for making recommendations "have the facts at their disposal."[61] Despite Ploscowe's request, Anslinger offered no meeting times.

Ploscowe responded immediately to Anslinger's letter asking for clarification of the "'glaring inaccuracies, manifest inconsistencies, apparent ambiguities, important omissions, and even false statements.'" Twice in his March 6, 1958, letter, Ploscowe requested a meeting with the commissioner.[62] The ABA-AMA Joint Committee attempted to maintain an amicable relationship with the FBN by continuing to request meetings and by granting Anslinger's requests for subsequent copies of the *Interim Report.* Despite his March 4 offer to meet, Anslinger continued his dogmatic response by ignoring repeated invitations to meet with the committee. Additionally, when his own report was published, he refused to provide the ABA-AMA Joint Committee with a copy. Dr. David F. Musto, who interviewed Anslinger after his retirement, described this as typical of Anslinger's behavior throughout his thirty-two-year tenure as commissioner. Musto said that Anslinger survived in Washington because "he created the agenda, attacking, rarely responding."[63]

No Meeting and No Report

In a July 11, 1958, letter, King described to Ploscowe the "silly business" of his futile attempts to procure a copy of the FBN report from both Anslinger's and Boggs' offices. He also told Ploscowe that the FBN had attempted "to get funds for printing from Russell Sage," the foundation that had granted funds to the ABA-AMA Joint Committee for its initial research and publishing costs.[64]

On that same day, King sent letters to Anslinger and Boggs again requesting a copy of the FBN report responding to the ABA-AMA Joint Committee's *Interim Report.* In both letters, King stressed three things: first, the joint committee's need for the FBN Advisory Committee's research results, particularly regarding "the factual misstatements and inaccuracies to which you have referred in prior correspondence"; second, the joint committee's continued desire to meet with Anslinger and the FBN Advisory Committee at the FBN's convenience; and third, the need "to correct one patent misunderstanding that has apparently arisen because of some of the present references to our *Interim Report.*" Correcting that "misunderstanding," King wrote, "None of us has ever been a proponent of the so-called clinic system of free distribution of drugs to addicts." In his letter to Boggs, King clarified that

> one of the research undertakings which we suggested was the operation of a small-scale public facility to experiment in the out-patient treatment of addicted persons, but we did not recommend any such facilities as a solution for the problem and, in fact, our report contains no recommendations at all beyond the pursuit of further research and study.

To Anslinger, King confirmed, "none of us is suggesting or proposing a revival of the so-called clinic system for uncontrolled distribution of drugs." King called such newspaper reports "simply inaccuracies of interpretation."[65]

In the spring of 1958, Anslinger asked Hale Boggs, as chair of the Advisory Committee to the Federal Bureau of Narcotics, to respond to the ABA-AMA Joint Committee's requests for a meeting. Boggs wrote to Ploscowe on May 26, 1958, saying that

> this subject is so important and many of the issues raised in your report are so inconsistent with the views and legislation which many of us have had a part in in the past that I am sure you realize that we want to have ample time to cover each and every one thoroughly.

Thus, speaking for the FBN, Boggs too put off the ABA-AMA Joint Committee's request for a meeting until the FBN Advisory Committee had time to prepare its "report on the various points and issues covered by your Joint Committee in your *Interim Report*." Boggs reminded Ploscowe that the joint committee "has spent two years preparing its report" and the FBN Advisory Committee "has been working on [its] report for only two months." The FBN Advisory Committee missed the point of the invitation: working together to examine the facts of drug traffic and use. Yet the FBN Advisory Committee chose to rebut the ABA-AMA Joint Committee's draft report with its own formal report, with no discussion and thus no opportunity for exchange of research, experiences, or ideas. Boggs indicated that the FBN Advisory Committee "as of now has prepared 100 pages to cover the various issues raised in your report." About setting a meeting, Boggs wrote, "As soon as the report is completed I shall send you a sufficient number of copies for study by your Committee. After that we shall be very happy to give consideration to a meeting with members of your committee."[66] None of this was true. The ABA-AMA Joint Committee had to continue to beg the FBN Advisory Committee for a single copy of its report long after it reputedly had been published, and no meeting was forthcoming. Correspondence among ABA-AMA Joint Committee members showed surprise and frustration at the FBN Advisory Committee's actions.

Further, in this letter, Boggs admonished the ABA-AMA Joint Committee for not giving "all points of view . . . adequate consideration" at a recent meeting in Bethesda, Maryland. Ironically, Boggs, a supporter of Anslinger, stated that "our country being what it is needs all sides of every issue thoroughly aired, . . . particularly . . . in the case of Narcotics." Boggs suggested that the ABA-AMA Joint Committee not "hold meetings and sponsor speeches on this subject giving only one side of the question."[67] The irony increased, considering that the meeting to which Boggs referred was the Symposium on the History of Narcotic Drug Addiction Problems held in Bethesda, Maryland, in March 1958. This meeting was prompted by Dr. Harris Isbell's suggestion that the National Institute of Mental Health "sponsor a symposium to which would be invited experts from all of the several professions and official agencies nationally and internationally concerned with narcotic drug addiction problems."[68] Symposium speakers included U.S. Assistant Attorney General William F.

Tompkins, with whom Anslinger co-authored the 1953 book *The Traffic in Narcotics,* and FBN senior agent Malachi Harney, both of whom spoke at length, championing the FBN's punitive approach to the narcotics problem. In the introduction to the published proceedings, the symposium planners acknowledged Anslinger's help "during the planning stages," but expressed regret that the commissioner "unfortunately was unable to attend the conference itself," thereby again avoiding direct communication with the ABA-AMA Joint Committee members, three of whom spoke at the symposium. Boggs erred in saying that both sides of the issue were not represented at the symposium.[69]

Boggs closed his letter with another ironic twist, stating that as the author of the "so-called 'Boggs Narcotic Law,'" he hoped that "by frankly and openly discussing and analyzing the issues relating to Narcotics we can arrive at constructive and forward-looking legislation and enforcement methods," yet his letter's purpose was to forestall any discussion.[70]

In a 1990 interview, Rufus King said of the Bethesda symposium,

> In 1958, there was a very liberal group in the National Institutes of Health—Drs. Bob Felix, Ken Chapman, Lawrence Kolb—and they organized a three- or four-day session to hear both sides of the issue. When the proceedings were published, Anslinger got his boss, the Secretary of the Treasury, to say a bad word or two about every one of them, even James Bennett, the Director of Prison Bureaus, who was of equal stature in Washington to Anslinger himself. Or they got a letter of censure in their personnel file.

King tapped his desk and repeated, "Every one of them got some kind of expression of displeasure for daring to raise the issue."[71]

By July 1958, Anslinger publicly denigrated the ABA-AMA Joint Committee's work, as reported in the *Washington Report on the Medical Sciences,* a Washington weekly newsletter. It stated that at an AMA conference in San Francisco, "members of the AMA House of Delegates received in their packets a document marked 'Confidential.' This was an *Interim Report* by [a] joint committee of bar and medical associations, indorsing legalized dispensing of narcotics to addicts for outpatient treatment." This general statement left readers to make their own interpretations, the worst being that addicts would receive a supply of drugs to take out of the clinic to sell. Further, the *Interim Report* did not "endorse" but recommended an outpatient experimental clinic in Washington, DC. The *Washington Report* continued, saying that the *Interim Report* had been "leaked to the press and now, in consequence, Narcotics Commissioner Harry J. Anslinger had made public a detailed rebuttal and commentary by his own advisory committee." The *Washington Report* said that "Commissioner Anslinger not only scoffs at [the joint committee's] conclusions but doubts their objectivity" and quoted Anslinger, saying, "In this instance it would appear that the conclusions of the committee were determined by the composition of its membership and for all practical purposes, their conclusions preceded the formation of the committee."[72] The antagonism would continue.

The members of the AMA who met in July 1958 in California did receive copies of the *Interim Report.* According to King, the document "was printed in temporary form as an unbound document which looked more like proofs than a finished tome. . . . Out of excess caution, because the report was not being submitted for final action, this makeshift version even bore a legend marking it confidential and not for general distribution."[73]

Anslinger wrote to King on July 28, 1958, again delaying meeting with the joint committee, saying, "A place and date of meeting should be agreed on after the final report has been adopted."[74]

FBN COMMENTS ON NARCOTIC DRUGS, *OR THE FBN ATTACKS THE* INTERIM REPORT

The FBN Advisory Committee's 186-page report *Comments on Narcotic Drugs,* which was twice the length of the *Interim Report,* was a compilation of old and new essays in no particular order, and the essays themselves followed no discernable organizational pattern. The contents page was a list of fourteen contributors' names interspersed with an occasional heading such as "report" or "survey" or "Honolulu." The contributors—doctors, judges, FBN employees, and one editorial writer/narcotics expert—also followed no particular organization in their respective pieces. Most surprisingly, none of the contributors responded directly to the ABA-AMA Joint Committee's report. Instead, many of the contributors either rambled or generalized from FBN anecdotes, mirroring the bureau's historical public address strategies of appealing to fear or pity. Most of the contributors began their pieces with ad hominem arguments, using some derogatory general reference to the members of the ABA-AMA Joint Committee. Some contributors even resurrected the *ad misericordiam* claims from the 1930s' "reefer madness" days. Anslinger contributed surprisingly little to the *Comments,* but his introduction set the tone for the entire report. An excerpt from Anslinger's introduction follows:

> When one examines the composition of the joint committee of the American Bar Association and the American Medical Association, one finds that the members are, almost without exception, individuals who had identified themselves with one panacea. These single minded individuals then emerged under what appeared to be the sponsorship of the ABA and the AMA. The public is conditioned to expect that ABA and AMA committees are oriented toward impartial deliberation, rather than propaganda. In this instance it would appear that the conclusions of the committee were determined by the composition of its membership, and for all practical purposes, their conclusions preceded the formation of the committee.

Following is an excerpt from one representative *Comments* contributor. George H. White, FBN district supervisor in San Francisco, began by writing, "The report itself is replete with mis-statements, ambiguities, contradictions and assumptions of facts not in evidence," language very similar in diction and tone to Anslinger's language in his March 4, 1958, letter to joint committee director Morris Ploscowe. Further, White called Ploscowe, who wrote the ABA-AMA Joint Committee report's introduction, "obviously biased and uninformed" and accused him of grandstanding and attempting to "'sell papers' by advocating a shocking revision of the controls which have been achieved through many years of the most careful consideration by a vast majority of all concerned with the problem."[75] White attempted to discredit Ploscowe by calling him a member of the "King-Kolb-Lindesmith party line" as though simply associating with King, Kolb, and Lindesmith was evidence of wrong thinking. He also accused Ploscowe of "attempting to find material to support" his hypothesis and of ignoring other evidence that supported the FBN's policies. White portrayed Judge Ploscowe as a dupe who vacillated between "wearing his medical smock" and "donning his judicial robes." Further, without referring to specific parts of the ABA-AMA *Interim Report,* White dismissed the entire report, saying, "There is scarcely a page of this voluminous document which is not susceptible to challenge." Like the language in most of the other statements in the FBN's *Comments,* White's language throughout his brief three-and-one-half page statement was general and accusatory. He called the *Interim Report* a "confusing" document full of "glaring examples of . . . slanted statements and conclusions."[76]

Finally, White closed by stating that "the disturbing thing about this recapitulation is that it purports to represent the policies of the American Medical Association and the American Bar Association"; therefore, the FBN Advisory Committee members' "time and trouble" in refuting the *Interim Report* were necessary "to expose this 'study' for the superficial, prejudiced and inaccurate document that it is."[77] That was his final sentence, and it was representative of most of the concluding comments of the other contributors to the FBN's *Comments.*[78]

White's claim that the ABA-AMA *Interim Report* "purports to represent the policies" of the ABA and AMA implied that the joint committee did not have the approval of their professional associations. Yet both the ABA and AMA Houses of Delegates voted to approve the *Interim Report* early in 1958.[79] Other contributors to the FBN's *Comments* used the same strategies that had served the FBN so well in the 1930s—they capitalized on pathos, using distorted statistics, shock tactics, and even the marijuana insanity claim from that decade.

Even the cover of the FBN *Comments* appeared deceptive. King claimed that Anslinger purposely tried to make the FBN *Comments* look like the joint committee's *Interim Report* (see Figure 4.1). According to King,

> Official U.S. Government publications are usually covered with white stock, or a standard gray or buff. Commissioner Anslinger's *Comments* came out in a shade of blue that closely approximated the paper used by the Joint Committee. And the size, title and type were calculated deception.[80]

In fact, the two covers not only looked similar, but the title of the FBN *Comments* (see Figure 4.2) looked more like a title for the *Interim Report* of the ABA-AMA Joint Committee.

NARCOTIC DRUGS

INTERIM REPORT

of

THE JOINT COMMITTEE OF THE AMERICAN BAR ASSOCIATION AND THE AMERICAN MEDICAL ASSOCIATION ON NARCOTIC DRUGS

AMERICAN BAR ASSOCIATION

RUFUS KING, ESQ., *Chairman*

JUDGE EDWARD J. DIMOCK

ABE FORTAS, ESQ.

AMERICAN MEDICAL ASSOCIATION

DR. ROBERT H. FELIX

DR. ISAAC STARR

C. JOSEPH STETLER, ESQ.

MORRIS PLOSCOWE

Director Narcotics Control Study

NEW YORK 1958

FIGURE 4.1. The cover of the *Interim Report of the Joint Committee of the American Bar Association and the American Medical Association on Narcotic Drugs,* 1958

Comments on

NARCOTIC DRUGS

INTERIM REPORT OF THE JOINT COMMITTEE

OF THE

AMERICAN BAR ASSOCIATION

AND THE

AMERICAN MEDICAL ASSOCIATION

ON

NARCOTIC DRUGS

BY

Advisory Committee to the Federal Bureau of Narcotics

U. S. TREASURY DEPARTMENT
BUREAU OF NARCOTICS
WASHINGTON, D. C

FIGURE 4.2. The similar cover of the FBN's *Comments* on the ABA-AMA joint committee report, 1958

1959 NBC MONITOR—*STILL NO DISCUSSION*

Between 1958, when the ABA-AMA Joint Committee's *Interim Report* was published, and 1961, when the joint committee's work was published in its entirety in *Drug Addiction: Crime or Disease?,* NBC radio's Walter McGraw, host of the *Monitor* talk show, devoted one weekend in 1959 to the serious question of whether "a drug addict is a dangerous dope fiend who preys on innocent citizens or a sick neurotic patient being preyed upon by the law enforcement agencies of our country."[81]

The *Monitor* forum was not a panel discussion but an eleven-part series of individual interviews with Commissioner Anslinger; Peter Terranova, narcotics inspector for the New York City Police Department; C. V. Narasemen, United Nations undersecretary for narcotics control; James V. Bennett, director of the Federal Prison Bureau; ABA-AMA Joint Committee members Rufus King and Dr. Isaac Starr; Dr. Lawrence Kolb, first director of the Federal Narcotics Hospital at Lexington, Kentucky; and Dr. Samuel Irwin of the Schering Chemical Company and formerly public health service officer at Lexington. This was as close as the ABA-AMA Joint Committee ever came to a discussion about the *Interim Report* with Anslinger, and although host Walter McGraw did ask the participants to respond to one another's statements, this forum allowed for no discussion among participants. They used the forum as a place unequivocally to air and to defend their positions. Throughout the forum Anslinger was unwilling to consider alternative viewpoints. He also maintained his own erroneous interpretation of the ABA-AMA Joint Committee's recommendations.

Joint committee member Rufus King began thc series by clarifying that "we are talking about adjusting the status of the addict and not in any sense reducing the penalties and the enforcement activities that are aimed at the [nonaddicted] peddler," whom King called "the most vicious predatory exploiter." He then explained that "frequently the addict and the peddler are identified together" because the addict has to become a peddler to earn enough money to support his habit.[82]

This was the critical distinction that the ABA-AMA Joint Committee made—that the nonaddicted peddler deserved severe legal punishment, but the addict and the addict-peddler were sick people needing medical attention. Yet as Dr. Lawrence Kolb pointed out, "Addicts have no place to go when they need help. They go to most doctors and get a rather cold welcome as you might suspect." Rather than the doctor being "at liberty to take into his office, as they can in England, an addict and treat him there," the doctor in the United States has been "terrorized by prosecutions, . . . a ridiculous situation."[83]

In response to the different interpretations, Anslinger attacked his opponents rather than respond to the issue. Making an attempt at humor, Anslinger suggested that the "proponents [of ambulatory treatment] should hire a good lawyer to read those laws and interpret Supreme Court decisions; they have not read it through very thoroughly. The idea that the interpretation is different is just utter nonsense."[84]

Anslinger was referring to interpretations of rulings in narcotics cases. In its work, the Narcotics Bureau used the combined language of two cases—*Webb v. United States* (1919) and *United States v. Berhman* (1922)—to write FBN Regulation Number 5, Article 167. This legislation included the following statement:

> An order purporting to be a prescription issued to an addict or habitual user of narcotics, not in the course of professional treatment but for the purpose of providing the user with narcotics sufficient to keep him comfortable by maintaining his customary use, is not a prescription within the meaning and intent of the [Harrison] act; and the person filling such an order, as well as the person issuing it, may be charged with violation of the law.[85]

The ABA-AMA Joint Committee cited more recent court decisions in its interpretation, specifically the 1936 case of *United States v. Anthony* in which a physician "was approached by the City of Los Angeles to take over the treatment of addicts who were former patients at the City's narcotics clinic, before it was closed." Dr. Anthony thoroughly examined the patients and prescribed drugs for them. At his hearing, two physicians testified against Dr. Anthony and three testified for him, agreeing, "such prescription was good professional practice."[86] In acquitting Dr. Anthony, the district court judge said, "Ultimately, the question to determine is not whether the judgment used was good or bad, but whether the defendant believed . . . that the treatment he administered was proper by ordinary medical standards."[87]

In his *Monitor* interview, Dr. Kolb conceded, "Of course, the law itself helped to reduce the number of addicts; there's no doubt." Yet he maintained that the interpretation was, in fact, "all wrong," and that this perpetuated the connection between addicts and criminal activity. Doctor arrests necessarily "drove these incurable cases into the illegal market, and that is where the crime is involved, the crime of getting these drugs regardless." Rather than the addict-crime connection perpetuated by the FBN, Kolb argued that the connection occurred because addicts "were trying to make a few dollars to get money to supply their own physical need." Otherwise, "the drug itself has exactly the opposite effect, and that's been proven here not only by me but other authorities. The addict is sedate, serene, and lethargic."[88]

Anslinger was always quick to point out that the United States had tried the ambulatory clinic approach in the 1920s and that it had failed; therefore, he concluded, it could not work. Rufus King described the 1920s system:

> Back in the beginning of the campaign to wipe out the narcotic traffic in a number of states and cities where there were a number of addicts the health authorities set up public clinics to administer drugs to addicted persons. Some of these clinics were very carefully and well administered. The addicts came, were registered, and were treated by responsible people within the clinics. Some of the clinics went haywire. Some of them became peddling stations. Some of them were corruptly administered, but about 1922 or '23, pressure from the federal enforcement authorities closed them all, and now every time anyone talks about clinic facilities, the cry that is raised is that they would become feeding stations for addicts.[89]

Anslinger occasionally used far-reaching analogies to make his point:

> The idea of a government poisoning its citizens with narcotics is nonsense. Why don't they set up bar rooms for alcoholics. . . . Why not furnish everybody with what they want, bullets, or department stores for kleptomaniacs, and so on. . . . We would be the only country in the world that has narcotic bar rooms for its citizens.[90]

Perhaps trying to elicit the listeners' fear, he added, "Communists would certainly like to see something like that come about." Rufus King responded to this analogy:

> The cry that is raised is that the clinics would become feeding stations for addicts, and one of the arguments is to the effect that you might as well open bars to treat drunks and on ad absurdum. This is a ridiculous distortion of the history of the clinics. In my opinion, a carefully administered public health facility to deal with the addict on the street is part of the important first step in treating this problem.[91]

Regarding this, Anslinger returned to the FBN policy of compulsory hospitalization for addicts, saying, "What we have been trying to bring about is not to penalize the addict but compulsory hospitalization in an enclosed institution because any other kind of a cure will not work." As proof of this policy's effectiveness, Anslinger offered, "Under present policy, which originated in 1914, . . . drug addiction in the United States was reduced from one in every 400 of the population to one in every 3,500." However, the commissioner also admitted that "we make no distinction between the addict-trafficker and the non-addict-trafficker," indicating that the addict-trafficker would probably face withdrawal in a jail cell rather than in a hospital.[92]

Anslinger made his own charge against the members of the joint committee, calling their "proposal to have the government supply thousands of teenagers with heroin, marihuana, and cocaine . . . slow murder that would make narcotic slaves of these people for the remainder of their lives." According to Anslinger, the joint committee's report

> violates treaties, federal laws, state laws, and Supreme Court decisions [but] it's a way for a proponent to get his name in the newspapers, and he loves to see his name. He always gets a little headline if he urges this sort of thing. The opponent is shunted to the want ads.[93]

In their closing statements in the *Monitor* series, both King and Anslinger maintained their polar positions, King proposing change and Anslinger resisting it. King discussed the "cruel" position in which current policy placed the addict. "If you gave that addicted person any other bargain on the harshest terms you could set forth, he would take it as an alternative to living in this shadow land at the mercy of his exploiter, the peddler." King recommended a series of options for addicts, beginning with making public health facilities available to them where they would "submit to an experimental withdrawal dose." Those addicts whose addiction was more severe could be withdrawn in a hospital. Finally,

> if at the end of all the efforts it's still indicated that a narcotic regime is necessary for him, then give it to him so that he could make the best adjustment he could to society under the care of medical authorities as an addict.[94]

Not surprisingly, Anslinger's closing remark refuted King's proposal for ambulatory treatment: "This idea of an addict going around to a doctor to get more morphine was tried many years ago, but it has failed." Each time Anslinger referred to the early 1920s experiment, he simply dismissed it without explanation. Anslinger closed with an analogy: "This addict is like a typhoid carrier; he will spread crime and disease wherever he goes. He will spread addiction," precisely what King said the ambulatory policy could prevent, but precisely what Anslinger knew the *Monitor* audience feared.[95]

LEGACY OF THE ABA-AMA JOINT COMMITTEE

The *Interim Report* and *Final Report* of the ABA-AMA Joint Committee were ultimately published together in 1961 under the title *Drug Addiction: Crime or Disease? Interim and Final Reports of the Joint Committee of the American Bar Association and the American Medical Association on Narcotic Drugs* by Indiana University Press. Lindesmith was a professor at Indiana University. According to King, a few years after the ABA-AMA Joint Committee concluded its work, "the American Bar Foundation [ABF] commissioned a staff member to evaluate what it had done—and this resulted in an ABF study which gave the Joint Committee yet another polite buffeting."[96]

The ABF addressed each of the three areas in which the ABA-AMA Joint Committee had made recommendations. King wrote:

> The staff reviewer found it "not readily apparent" what Judge Ploscowe had been talking about when he referred to an experimental clinic, and found it "difficult to determine" what the Joint Committee intended. He concluded that because any experimental drug program would involve "complex medical-legal questions," might raise ethical problems, and—Judge Ploscowe to the contrary notwithstanding—would conflict with present laws and treaty obligations, the suggestion ought to be shelved until the way was better paved by research and educational efforts.[97]

The joint committee's views about the value of educational and preventive measures fared slightly better in this appraisal, being tentatively accepted, although the report noted that Judge Ploscowe should have been more specific and predicted that not much would really be accomplished along such lines. In the matter of legal research, the ABF evaluation concluded that Judge Ploscowe and the joint committee had not understood each other, and that although some of the problems were "discussed fully and competently by Judge Ploscowe," tackling them would require large expenditures of time and money and would probably not produce meaningful results.[98]

Finally, the joint committee's dissatisfaction with too-severe criminal penalties was dismissed as starting from "an unproved assumption," and it was concluded, "evidence has not been marshaled to demonstrate that severe penalties are or are not the answer to the narcotics problem of the United States." From this it followed that further studies would be "both presumptive and unpropitious," and so the joint committee's discussions in this area were put aside with a near-scolding: "Any proposal for sweeping revision of legislation in this field should reflect the most careful consideration of all possible factors and be supported by the most complete evidence of the need for change."[99]

In his book *The Drug Hang-Up,* King recalled this episode in a chapter titled, "ABA-AMA, No Match for HJA." King was asked in a 1990 interview, "Why was the ABA-AMA no match for Anslinger?" King responded, "One of the worst mistakes I ever made in my life was the ABA-AMA Committee," and he gave two reasons: "Anslinger's penchant for retribution and Anslinger's connections in Congress." About Anslinger's penchant for retribution, King referred to his earlier example of the retribution that each of the participants at the Bethesda Symposium of 1958 had received. About Anslinger's connections in Congress, King said,

> Anslinger had a whole stable up there [in Congress]. Anslinger had done favors for people with trouble I have no idea about, but I'm willing to speculate that with a force of 300 agents and officers in every major city there were a lot of favors. If he learned that a Congressman or Congressman's friend was going skiing someplace, there would be a Treasury agent meeting him at the airport. That was done a lot. That's the name of the game. But there was that kind of polishing, and then a great flow of information and assistance for anybody who wanted to pick up the Bureau line and give speeches and that sort of thing. His biggest puppet up there was Hale Boggs on the House side. Hale Boggs made something of a career of being tough on dope fiends.[100]

Yet King added, "In my personal contact with him [Anslinger], I never had any feeling of insincerity. I didn't get that close to Anslinger, but I never had the feeling that he had any doubts about what he was doing."[101]

Perhaps King was right: the ABA and AMA were no match for Anslinger. The "full and free discussion" that Lindesmith called for never happened. How did Anslinger and the FBN render these powerful bodies almost impotent?

CONCLUSION

Throughout his lengthy career, Anslinger relentlessly sought to put an end to the drug traffic in the United States. He was equally relentless in pursuing what he felt was the most logical corrective action for the problem: strict punitive measures in the form of fines and lengthy prison sentences. The legislation enacted during those three decades—The Marihuana Tax Act of 1937, the Boggs Act of 1951, and the Narcotic Control Act of 1956—demonstrated the extent of Anslinger's belief in increased punishment as a deterrent to drug trafficking, peddling, and using.

For almost all his public life, Anslinger remained unchanged on at least six points:

1. *Narcotics education would produce more harm than good.* As late as his final appearance as commissioner before the House Appropriations Subcommittee on January 30, 1962, Anslinger testified, "The education route has not worked at all. No education program has done any good, because New York, with all these people taking a compulsory course in narcotics, you still get an increase."[102]

2. *Increased punishment was the best single deterrent to crime.* When Anslinger testified before the House Appropriations Subcommittee on January 23, 1958, two years after enactment of the Narcotic Control Act, he referred to the act, saying, "I don't think you could ask for anything better."[103] The following year, when New Jersey Representative Gordon Canfield asked Anslinger if he recalled a movement in 1958 to repeal the mandatory sentences, Anslinger responded, "I had cold chills," and added that such a move would have set narcotics law enforcement "back twenty years."[104] Finally, in his last appearance, on January 30, 1962, Anslinger stated directly, "The trafficker does not get what he should," advocating still heavier penalties.[105]

3. *Marijuana was the gateway drug.* While Anslinger's comments on marijuana's addictiveness wavered depending on his goal at a particular moment, he never wavered from his notion that it was the stepping-stone drug that led users to more dangerous narcotics. On January 27, 1959, four years before the end of his tenure, he testified that marijuana "is a habit-forming drug, but not an addiction-forming drug, although it is a steppingstone to an addiction-forming drug, but it is a very dangerous drug and is so regarded all over the world."[106] The debate on medical marijuana had—perhaps mercifully—not yet begun.

4. *Marijuana could produce unpredictable effects in users.* While for part of his career Anslinger suggested that the connection between marijuana and crime might have been exaggerated, his 1961 book, *The Murderers,* ended with an appendix titled "Main Narcotic Drugs and Their Derivatives." Under "Cannabis," Anslinger and his co-author Will Oursler wrote, "Effects vary with the individual and the strength of the drug. All varieties may lead to acts of violence, extremes, madness, homicide."[107] Further, one of his last public comments on that subject reinforced his earliest notion. Eight years after his retirement, when Anslinger was residing near his hometown of Hollidaysburg, he responded to newspaper coverage of a speech given

locally by Dr. Robert G. Shaheen, who "recommended more permissiveness in the use of marijuana." In a March 18, 1970, letter to the editor of *Blair Press,* Anslinger refuted Dr. Shaheen, writing, "I cannot allow Dr. Shaheen's misleading talk before the Rotary Club which has been made public to go unanswered."[108] Anslinger explained, "marijuana distorts time and distance and accounts for considerable slaughter on the highways, even with the smoking of one cigarette." With no explanation of his figures, Anslinger claimed "there are some 30,000 deaths on the highways due to drunken driving. If Marijuana were legalized, there would be 100,000 deaths a year; a situation worse than Vietnam." Connecting the dangers of marijuana use to young people, Anslinger wrote, "Dr. Shaheen's recommendation for permissiveness would cause widespread physical and moral deterioration in our youth." Finally, the commissioner quoted the United Nations Economic and Social Council's position on marijuana. This statement recognized "that cannabis is known *inter alia* to distort perception of time and space, modify mood and impair judgment, which may result in unpredictable behavior, violence and adverse effects on health, and that it may be associated with the abuse of other drugs such as LSD, stimulants and heroin."

5. *A young person's upbringing was instrumental in whether that person became involved with drugs.* In a commencement speech at Saint Francis College in Loretto, Pennsylvania, on June 5, 1966, Anslinger condemned those college and university administrators who allowed drug use on their campuses. But he said to the Saint Francis graduates, "You have going for you that volatile but resurgent and possibly indomitable instrument: the human spirit. . . . Jerks and beatniks multiply but conceivably the squares may yet contrive to clobber the delinquents."[109]

6. *Ambulatory treatment would not work.* In an April 28, 1962, *New York Forum* broadcast over WCBS-TV, Anslinger pointed to the failed ambulatory clinic experiment of the 1920s, saying that those clinics "were all closed by actions of the legislatures and by resolution of the American Medical Association."[110] Anslinger never mentioned that legislators and members of the AMA agreed that the system failed because not enough thought and preparation went into it; the system was not well controlled, and too few doctors were trying to treat too many patients.

Anslinger maintained these themes with little change over his thirty-two years as commissioner and in his public statements after his retirement. His adamant persistence in these public arguments contributed to his credibility, particularly over such a long period. At times, however, Anslinger was factually mistaken, yet seldom did anyone question him. What, then, made him an expert in his field? He enjoyed several advantages that contributed to his expert credibility.

One advantage was that Anslinger was the first on the scene. As the original commissioner of the Federal Bureau of Narcotics, he had no predecessor. No previously established precedent was in place for the limits of the commissioner's authority.[111] No one before him held a position that involved speaking publicly on the issue of drugs, and Anslinger relished speaking and writing about narcotics. Simply by virtue of the appointment, by being labeled commissioner, Anslinger received credibility. In the eyes of almost everyone, Anslinger was an expert, but he was an expert with benefit-of-the-doubt credibility. One can maintain this kind of credibility until refuted by an expert perceived to be equally or more credible, and that did not occur in Anslinger's thirty-two years.

A second advantage that Anslinger enjoyed was the poverty of his field. Before the FBN's inception, the medical and law enforcement professions knew relatively little about narcotic drugs and their effects. Even during Anslinger's tenure, researchers undertook few studies, and those were generally questionable in terms of control, numbers of subjects, selection of subjects, types and amounts of drugs administered, and methods of measuring emotional response. During these years, researchers debated but did not concur on what exactly drug studies should attempt to measure: effects of the drugs themselves or what conditions cause people to become involved with drugs, among others. With relatively little concrete scientific information in the public domain, Anslinger authoritatively could state his position and few could successfully refute him.

Anslinger's third advantage, a reality until the 1970s, was a general lack of public interest in the subject of illegal drugs. Those who expressed the most interest in drug-related matters were public officials who used the drug issue as part of their election campaigns, such as Senator Price Daniel of Texas in his gubernatorial quest. What is remarkable is that Anslinger maintained his power despite others' seeming lack of interest in drug issues.

His fourth advantage was that Anslinger's listeners were predisposed to listen to him and accept his opinions. In addition to legislators and the law enforcement community, Anslinger had a select group of followers. Certainly wholesale and retail pharmaceutical firms and pharmacists not only listened to the commissioner but frequently sought him out to speak at their conventions, obviously interested in his opinions, perhaps to some extent for economic reasons. Members of the medical profession out of necessity listened to Anslinger, probably because of the close surveillance that the commissioner instructed FBN agents to keep on doctors' offices. In addition, Anslinger was generally popular with civic and philanthropic groups who, as one of their foci, were concerned with young people and the conditions of the country in which they were growing up. Comparatively speaking, the number of the commissioner's followers was quite small, and the population at large never seemed overly interested in drug issues during the Anslinger years, yet the followers that he did have listened carefully to him and respected his views.

A fifth advantage that Anslinger had was his all-consuming zeal for his work. Some opponents of his policies suggest that Anslinger did not believe some of the arguments he advanced to promote the policies he desired. In an interview, Rufus King suggested that Anslinger created a marijuana "bugaboo" to legitimize longer jail sentences for these drugs' distribution and use. However, senior agent Garland Williams, responding to the question, "Do you think Anslinger did that to scare people, or did he really believe that marijuana was bad?" answered, "A marijuana user goes crazy when he's under heavy doses. . . . Anslinger emphasized in some of his writings that the marijuana user is dangerous because it was the truth. Even the other kind of drug users called them 'weed heads' because they're crazy, they're dangerous."[112] What Anslinger actually believed on this may never be known, particularly given a letter in his files. The November 11, 1948, letter from Philip H. Glaessner Garcy alluded to a 1938 cruise on which the commissioner " 'debunked' the version of the time, of the tremendous action of the 'weed' upon the nervous system."[113] Yet Williams' account of Anslinger's belief may have been the assumption under which the FBN agents operated.

Anslinger's zeal gave him an advantage because his fervor contributed to the sincerity and authenticity of his arguments. However, a zealot can also be dangerous and can misuse authority.

Although Anslinger's career is a mixed success story, his major legacy lives on. He left the public debate on how to deal with illicit drugs the same way he entered it, adamant in his position that the United States' drug problem could best be solved in the criminal domain, not in the public health domain, a position he held for his entire life. Despite his foibles and idiosyncrasies, Anslinger initiated policies that remain relatively intact today. As the enactment of these three major pieces of narcotics legislation shows, Anslinger dramatically increased the United States' intolerance for narcotics, narcotics sellers, and addicts. He also increased the United States' appetite for stronger laws with heavier punishment to control illegal drugs. This appetite has not yet been satisfied.

URLs FOR SELECTED PRIMARY DOCUMENTS

The Traffic in Narcotics by H. J. Anslinger and William F. Tompkins
<http://www.druglibrary.org/schaffer/people/anslinger/traffic/default.htm>

Drug Addiction, Crime or Disease? Interim and Final Reports of the American Bar Association and the American Medical Association on Narcotic Drugs, 1961
<http://www.druglibrary.org/schaffer/Library/studies/dacd/Default.htm>

Legal References on Drug Policy; Federal Court Decisions on Drugs by Decade, 1950
<http://www.druglibrary.org/schaffer/legal/legal1950.htm>

Legal References on Drug Policy; Federal Court Decisions on Drugs by Decade, 1960
<http://www.druglibrary.org/schaffer/legal/legal1960.htm>

NOTES

1. Harry J. Anslinger and William F. Tompkins, *The Traffic in Narcotics* (New York: Funk and Wagnalls Co., 1953), p. 166. Also published in Harry J. Anslinger,

"The Sheriff and Narcotic Enforcement," *International Criminal Police Review* (December 1952): 318.

2. Anslinger and Tompkins, *Traffic in Narcotics,* p. 166; Anslinger, "The Sheriff and Narcotic Enforcement," p. 321.

3. "Anslinger Calls for Stiffer Penalties on Dope Peddlers to Protect Youth of Nation," *Altoona Mirror,* November 23, 1953. AP, Box 5, File 5. Because Altoona is near Anslinger's hometown of Hollidaysburg, the *Mirror* covered his statements closely. Anslinger moved to Altoona after his retirement and lived there until his death.

4. "Anslinger Asks Senate Action on Addict Bill; Hospitalization Law Still Needed, He Says; Drug Use Declines," *Washington Post,* November 23, 1953. See also Harry J. Anslinger, "Extent of Narcotic Addiction in Youths Under 21," Senate Sub-Committee to Investigate Juvenile Delinquency, November 1953, Anslinger files, Drug Enforcement Administration Library, Washington, DC.

5. "Anslinger Calls for Stiffer Penalties," *Altoona Mirror,* November 23, 1953.

6. Alfred R. Lindesmith, "Traffic in Dope, Medical Problem," *Nation,* April 21, 1956, p. 339.

7. Ibid.

8. See "Study of Narcotics Problem in the United States," *Congressional Record,* 84th Congress, 1st Session, February 21, 1955, in which Senator Price Daniel introduced Senate Resolution 60, proposing "that the Committee on the Judiciary, or any duly authorized subcommittee thereof, is authorized and directed to conduct a full and complete study of the narcotics problem in the United States, including ways and means of improving the Federal Criminal Code and other laws and enforcement procedures dealing with the possession, sale, and transportation of narcotics, marihuana, and similar drugs." On March 18, 1955, Senator Lyndon B. Johnson of Texas entered Senate Resolution 67, "Study of Narcotics Problem in the United States."

9. *Congressional Record,* 84th Congress, 1st Session, May 31, 1955, 101:7231.

10. Hearings Before the Subcommittee on Improvements in the Federal Criminal Code, U.S. Senate, *Illicit Narcotics Traffic,* 84th Congress, 1st Session, June 2, 1955, p. 9. See also AP, Box 1, File 8, "Speeches by Harry J. Anslinger 1938-1969."

11. *Illicit Narcotics Traffic,* pp. 16-17.

12. Ibid., p. 44.

13. Ibid., p. 45.

14. Ibid.

15. "Will Tough Laws Curb Dope?" *Daily News,* July 30, 1956. AP, Box 5, File 6. In this article, an unnamed senator who called for a "fresh approach on the problem" accused Anslinger of "resenting the fact that we invited a doctor favoring the English system to testify at the Senate subcommittee hearings and describe the English program."

16. *Illicit Narcotics Traffic,* pp. 63, 78.

17. Anslinger continued to cite this figure in subsequent years. For example, during a 1959 NBC radio show, *Monitor,* Anslinger offered this same explanation, claiming that all addicts came to the FBN's attention within two years of their becoming involved with drugs.

18. *Illicit Narcotics Traffic,* pp. 15-16.

19. Alden Stevens, "Make Dope Legal," *Harper's Magazine,* November 1952, p. 41.

20. Hearings Before the Subcommittee of the Committee on Appropriations, House of Representatives, *Treasury Department Appropriation Bill for 1941,* 76th Congress, 3rd Session, December 14, 1939, p. 431.

21. Lindesmith, "Traffic in Dope," p. 337.

22. "Narcotics Users Are Put at 60,000," *The New York Times,* June 3, 1955, p. 25.

23. *Congressional Record,* 84th Congress, 1st Session, June 9, 1955.

24. *Congressional Record,* 84th Congress, 1st Session, June 20, 1955, 101:8776.

25. *Congressional Record,* 84th Congress, 1st Session, July 27, 1955, 101: 11683.

26. *Congressional Record,* 84th Congress, 1st Session, July 21, 1955, 101: 11138.

27. "Stiffer Law Asked by Narcotics Chief," *The New York Times,* September 18, 1955, p. 74. Copyright © 1955 by The New York Times Co. Reprinted with permission.

28. Lindesmith, "Traffic in Dope," p. 339.

29. "Free Narcotics to Be Pondered," *The New York Times,* September 17, 1955, p. 36.

30. Charles Grutzner, "Doctors Opposed on Aid to Addicts: Academy Plan for Supplying Narcotics to Incurables Is Condemned at Hearing," *The New York Times,* September 21, 1955, p. 35. Copyright © 1955 by The New York Times Co. Reprinted with permission.

31. *Illicit Narcotics Traffic,* p. 1367.

32. Grutzner, "Doctors Opposed on Aid to Addicts," p. 35. Copyright © 1955 by The New York Times Co. Reprinted with permission.

33. *Illicit Narcotics Traffic,* p. 1378.

34. Ibid., p. 1379.

35. Grutzner, "Doctors Opposed on Aid to Addicts," p. 35. Copyright © 1955 by The New York Times Co. Reprinted with permission.

36. "House Unit Maps Tightened Penalties, but Bars Death Decree Asked in Senate," *The New York Times,* October 24, 1955, p. 29. Remarkably, Daniel added that the increased penalties requested for all narcotics included marijuana. More specifically, Daniel stated, "The Committee recommends that the penalties for violating the marihuana laws be raised considerably. I would be willing to include marihuana sales, along with heroin, as being punishable with the death penalty." *Congressional Record,* 84th Congress, 2nd Session, January 9, 1956, 102:273.

37. "House Unit Maps Tightened Penalties," p. 29.

38. "Inquiry Will Link Narcotics, Crime: Senate Group Is to Report Tomorrow to Congress on Tripling of Addicts," *The New York Times,* January 8, 1956, p. 54.

39. "Green Light for Daniel," *Time,* April 9, 1956, p. 36. *Time* reported that Senator Daniel "feared that the Texans who sent him to the Senate for six years might be peeved if he applied after four years for another position." So he let the voters decide. A week before the *Time* article appeared, Daniel asked his constituency in a radio address whether they wanted him to run for governor, and "29,000 of them sent him messages urging him to run."

40. "Roundup Time," *Time,* August 6, 1956, p. 18.

41. *Congressional Record,* 84th Congress, 2nd Session, January 9, 1956, 102: 273.

42. *Congressional Record,* 84th Congress, 2nd Session, April 30, 1956, 102:71-72. See also testimony on Senate Bill 3760, "A Bill to Provide for a More Effective Control of Narcotic Drugs," *Narcotic Control Act of 1956,* 84th Congress, 2nd Session, May 4, 1956, 1-44. Anslinger did not testify on May 4, 1956.

43. "'New Tack' Urged in Narcotic Fight: Bill Asks Stiff Penalties," *The New York Times,* May 15, 1956, p. 27.

44. *Narcotic Control Act of 1956,* 84th Congress, 2nd Session, June 19, 1956, House Report Number 2388, p. 65.

45. Hearings Before the Subcommittee of the Committee on Appropriations, House of Representatives, *Treasury-Post Office Departments Appropriations for 1958,* 85th Congress, 1st Session, February 4, 1957, p. 349. In this statement, Representative Canfield addressed the commissioner as "Dr. Anslinger." Although he never earned a doctoral degree, Anslinger did receive two honorary doctorates: an honorary doctorate of laws from the University of Maryland in 1955, and an honorary doctorate of humanities from Saint Francis College, Loretto, Pennsylvania, in 1966. He also received from his alma mater, The Pennsylvania State University, a distinguished alumni award; that university does not give honorary degrees.

46. *Drug Addiction: Crime or Disease? Interim and Final Reports of the Joint Committee of the American Bar Association and the American Medical Association on Narcotic Drugs* (Bloomington: Indiana University Press, 1961), p. 6. On page x of the introduction to this document, Alfred R. Lindesmith explained "only a limited number of copies of this document [the *ABA-AMA Interim Report*] were printed, primarily for the use of the Committee itself and for circulation in the Houses of Delegates of the two associations. A few additional copies were printed and sold to persons who requested them." In 1961, the *Interim Report* and *Final Report* were published together in one document with an introduction and appendixes. In the following endnotes, excerpts from this publication will be referred to as *ABA-AMA Interim Report* or *ABA-AMA Final Report* or *Drug Addiction: Crime or Disease?*

47. *ABA-AMA Interim Report,* p. 7.

48. Ibid., p. 9. Choosing Judge Ploscowe rather than a medical doctor did not mean that the committee was minimizing the need for medical research on the drug issue. Instead, the committee felt that the recent report of the AMA's Council on Mental Health on Narcotic Addiction was so thorough that little further medical research was necessary and that the joint committee could concentrate on the "legal, administrative and sociological aspects" of the problem.

49. *ABA-AMA Interim Report,* p. 9.

50. Ibid., p. 7.

51. Ibid., p. 14.

52. Ibid.

53. Advisory Committee to the Federal Bureau of Narcotics, *Comments on Narcotic Drugs: ABA-AMA Interim Report of the Joint Committee of the American Bar Association and the American Medical Association on Narcotic Drugs* (Washington, DC: U.S. Government Printing Office, 1958), p. viii. This publication will be referred to subsequently as *FBN Comments.* Anslinger's writing is noticeably absent from this publication, with the exception of the front matter. This consisted of a

brief history of how the publication was created, a list of joint committee members, a list of advisory committee members, and brief statements on narcotic drugs from the National Academy of Sciences and the United Nations. Members of the advisory committee wrote the remaining 185 pages. However, Anslinger may have put together the final two entries in the *FBN Comments,* a twenty-page section titled "General Comments" and fourteen pages of "Addiction Charts." The "General Comments" section cites excerpts in order of page number from the *ABA-AMA Interim Report* and refutes them.

54. Rufus King, *The Drug Hang-Up: America's Fifty-Year Folly* (Springfield, IL: Charles C Thomas, 1972), p. 165.

55. *Drug Addiction: Crime or Disease?,* p. viii.

56. Ibid.

57. Ibid., p. xiv.

58. *ABA-AMA Interim Report,* pp. 11-13. The second through fifth recommendations called for various research projects. In the second recommendation, the committee stated that a large-scale "study of relapse and causative factors . . . is absolutely indispensable for either the continuation of present policy or the formation of new policy." The third recommendation called for educational and preventative research. The committee noted, "the dissemination of accurate information about narcotic addiction has been neglected and even discouraged by some enforcement authorities." The committee suggested careful study "to determine whether a campaign of enlightenment might not produce good results." Fourth, the committee called for "legal research" to clarify ambiguous statutes and to evaluate "present legal approaches to narcotic addiction." Such research would have a twofold result: "better methods of dealing with the addict and more realistic and sounder means for controlling the illicit drug traffic." Finally, the fifth recommendation called for "research in the administration of present laws" to clarify the "considerable uncertainty and confusion in the enforcement of existing drug laws." The committee felt that a careful analysis of current application of narcotics laws would prove invaluable in establishing a "rational drug control program in this country."

59. Many of the studies cited in the *ABA-AMA Interim Report* are discussed in my dissertation: Rebecca Carroll, "A Rhetorical Biography of Harry J. Anslinger, Commissioner of the Federal Bureau of Narcotics, 1930-1962," doctoral dissertation, University of Pittsburgh, 1991.

60. Morris Ploscowe to Harry J. Anslinger, February 24, 1958, Rufus King Papers, housed in the Drug Policy Foundation, Washington, DC, hereafter referred to as RKP. None of this correspondence is contained in Anslinger's papers; however, this February 24, 1958, letter and Anslinger's March 4, 1958, response (see note 61) were published as part of the front matter in the *FBN Comments.*

61. Harry J. Anslinger to Morris Ploscowe, March 4, 1958, RKP. This accusation eventually reached much further than a letter to a single recipient. In addition to appearing as front matter in the *FBN Comments,* Anslinger used this exact phrase publicly on an NBC radio talk show in 1959, which is discussed later in this chapter.

62. Morris Ploscowe to Harry J. Anslinger, March 6, 1958, RKP.

63. David F. Musto, telephone interview with author, August 6, 1996.

64. Rufus King to Morris Ploscowe, July 11, 1958, RKP. The Russell Sage Foundation did not award the joint committee its second round of funding, probably because the *FBN Comments* intimated that the foundation, by its funding of the joint

committee's report, supported teenage drug use. "Directly and indirectly Russell Sage trustees were approached from the Treasury Department—the same Treasury Department which of course holds life-and-death power over all foundations through its discretion in granting tax exemptions—and given to understand that they were sponsoring a 'controversial' study, that the ABA and AMA spokesmen were irresponsible if nothing worse, and, in short, that it would be discreet to drop the project." King did not fault Russell Sage for dropping the project as "their foundation had other important works underway and the serious disfavor of Treasury could have jeopardized everything." King, *The Drug Hang-up,* p. 163.

65. Rufus King to Harry J. Anslinger, July 11, 1958, RKP. Rufus King to Hale Boggs, July 11, 1958, RKP.

66. Hale Boggs to Morris Ploscowe, May 26, 1958, RKP.

67. Ibid.

68. Robert B. Livingston (ed.), *Narcotic Drug Addiction Problems: Proceedings of the Symposium on the History of Narcotic Drug Addiction Problems* (Bethesda, MD: National Institutes of Health, 1958), p. viii.

69. The symposium proceedings is a 200-page document that will not be analyzed in detail here because representatives of both sides of the addict-crime issue made prepared statements that are the same as those presented in the joint committee's *ABA-AMA Interim Report;* the FBN Advisory Committee's *FBN Comments,* which claimed to respond to the *ABA-AMA Interim Report;* and the NBC Radio *Monitor* program to be discussed later and which contained interviews with Anslinger.

70. Hale Boggs to Morris Ploscowe, May 26, 1958, RKP.

71. Rufus King, interview with author, Washington, DC, January 11, 1990. Also in this interview, King said, "Incidentally, at that time, there was something that I miss now years later. At least then Anslinger sent Mal Harney and others from the Narcotics Bureau and guys with guns and badges. They were there, and they stood up, and they shouted at us, and we shouted back, and to some extent, Anslinger honored the sincerity of this group to stand up to us. And now these people, Bill Bennett and Reggie Walton and the rest of them, they won't debate. They make speeches and they call us names, but they won't appear and really defend themselves. To that end, Anslinger was a straighter shooter than the drug warriors now."

72. Gerald G. Gross (ed.), *Washington Report on the Medical Sciences,* July 7, 1958.

73. King, *The Drug Hang-Up,* p. 164. In additon, in Lindesmith's 1960 introduction to *Drug Addiction: Crime or Disease?,* which contained the *Interim* and *Final Reports,* he expressed concern about the FBN's negative reaction to the ABA-AMA Joint Committee's work: "The reader may be puzzled by the fact that the Bureau of Narcotics reacted negatively to the work of the Joint Committee. The Bureau is not assailed in the report and is mentioned in only a few places. The language of the report is calm, restrained, objective, and undogmatic. The main emphasis of the recommendations is upon the need for more reliable data, and the experimental project to explore the effects of providing addicts with legal drugs is cautiously stated. If this idea is as dangerous and unsound as its critics contend, the project proposed by the Joint Committee would go a long way toward discrediting it once and for all. This report of the Joint Committee neither states nor implies anything of a deroga-

tory nature concerning the purely law enforcement activities of the Bureau of Narcotics or of the police in general." *Drug Addiction: Crime or Disease?*, xi.

74. Harry J. Anslinger to Rufus King, July 28, 1958, RKP.

75. *FBN Comments,* p. 120.

76. White's phrase, the "King-Kolb-Lindesmith party line," referred to attorney Rufus King, joint committee member; Dr. Lawrence Kolb, first director of the Federal Narcotics Hospital in Lexington, Kentucky; and Lindesmith, who wrote the introduction to *Drug Addiction: Crime or Disease?* All three of these people dedicated a large part of their professional lives to reform of narcotics laws. About Lindesmith, King wrote, "At this writing one combatant of long standing remains very much in the fight, Dr. Alfred R. Lindesmith, whose doctorate is in sociology and whose unassailable base has long been increasingly important faculty posts at Indiana University. Lindesmith first occupied himself with the drug scene in the early thirties, bringing to his studies a Phi Beta Kappa background, a mixture of fine scholarship and articulateness, and a no-nonsense determination to observe carefully and talk about what he saw. Lindesmith's own books, *Opiate Addiction* (1947) and *The Addict and the Law* (1965), are landmark statements, while thanks to his good offices the Indiana University Press was for many years the main—and sometimes the only—publishing outlet for outspoken critics of official Narcotics Bureau dogma." King, *The Drug Hang-Up,* p. 62.

77. *FBN Comments,* p. 120.

78. For additional discussion of contributions to the *FBN Comments* as well as discussion of other public debates that Anslinger managed (e.g., New York Mayor Fiorello LaGuardia), see Carroll, "A Rhetorical Biography of Harry J. Anslinger."

79. King, *The Drug Hang-Up,* p. 170.

80. Ibid.

81. "Our Treatment of the Narcotics Addict," host Walter McGraw, *Monitor,* NBC Radio, New York, 1959, hereafter referred to as *Monitor.*

82. *Monitor,* Part One.

83. Ibid.

84. Ibid.

85. *ABA-AMA Interim Report,* p. 81.

86. Ibid., p. 80.

87. Ibid., p. 81. The definition and timing of "good faith" became a new demon for those physicians who continued to minister to addicts. According to Judge Ploscowe, on the surface, "a physician who treats and/or prescribes drugs for an addict patient in good faith according to medical standards will be protected from a conviction." However, a doctor's "good faith and adherence to medical standards can only be determined *after a trial,*" so the doctor has no protection before prescribing for an addict, "no way of knowing *before* he attempts to treat, and/or prescribe drugs to an addict, whether his activities will be condemned or condoned." Judge Ploscowe urged doctors to become active in their own governance and "not leave the task of determining good faith and proper medical standards to an ex post facto judgment made by twelve laymen on a jury." Ploscowe recommended "the profession itself, through its authoritative body, the American Medical Association, should lay down the criteria by which a physician's treatment of an addict can be judged." *ABA-AMA Interim Report,* p. 78.

88. *Monitor,* Part One.

89. *Monitor,* Part Eight.
90. Ibid.
91. Ibid.
92. Ibid.
93. Ibid.
94. *Monitor,* Part Eleven.
95. Ibid.
96. King, *The Drug Hang-Up,* p. 174.
97. Ibid.
98. Ibid.
99. Ibid., pp. 174-175.
100. Rufus King, interview with the author, Washington, DC, January 11, 1999. In addition to being the namesake for the Boggs Act of 1951, in the *Congressional Record,* Hale Boggs introduced the 1956 Narcotic Control Act (NCA) as the "Boggs-Daniel Narcotic Control Act." Anslinger referred to the NCA that same way in an article he published in the *FBI Law Enforcement Bulletin* in October 1962 (pp. 7-10); I could find no other references to the NCA by that name.
101. Rufus King, interview with author, Washington, DC, January 11, 1990. In addition, socially, Anslinger was perceived as an intelligent, charming person. He was well read, traveled the world, spoke several languages, and was described by one social acquaintance as "a very nice man." Anslinger was devoted to his wife, Martha Denniston, and her son Joseph. Interview with Rosamond Tirana, Washington, DC, January 10, 1990.
102. Hearings Before the Subcommittee of the Committee on Appropriations, House of Representatives, *Treasury-Post Office Departments and Executive Office Appropriations for 1963,* 87th Congress, 2nd Session, January 30, 1962, p. 289.
103. Hearings Before the Subcommittee of the Committee on Appropriations, House of Representatives, *Treasury-Post Office Departments Appropriations for 1959,* 85th Congress, 2nd Session, January 23, 1958, p. 103.
104. Hearings Before the Subcommittee of the Committee on Appropriations, House of Representatives, *Treasury-Post Office Departments Appropriations for 1960,* 86th Congress, 1st Session, January 27, 1959, p. 127.
105. Hearings Before a Subcommittee of the Committee on Appropriations, House of Representatives, *Treasury-Post Office Departments and Executive Office Appropriations for 1963,* 87th Congress, 2nd Session, January 30, 1962, p. 297.
106. *Treasury-Post Office Departments Appropriations for 1960,* p. 133. Marijuana, like cigarettes and alcohol, is a "starting" drug. However, according to David Courtwright, Herman Joseph, and Don Des Jarlais in *Addicts Who Survived: An Oral History of Narcotic Use in America, 1923-1965,* "Most marijuana smokers, if they went on to other drugs, chose pills, hallucinogens, or stimulants. It was in the ghetto that the ultimate graduation to heroin was still mostly likely to occur" (Knoxville: University of Tennessee Press, 1989), p. 352. In addition, most of the recovering addicts I interviewed stated that because the U.S. government so clearly lied about the effects of marijuana, it surely must have been lying about the effects of other narcotics—one of the reasons they advanced to stronger narcotics.
107. Harry J. Anslinger and Will Oursler, *The Murderers: The Story of the Narcotic Gangs* (New York: Farrar, Straus and Cudahy, 1961), p. 304.
108. Harry J. Anslinger, Letter to the Editor, *Blair Press,* March 18, 1970.

109. Harry J. Anslinger, "Commencement Address," Saint Francis College, Loretto, PA, June 5, 1966. DEA Library. At this ceremony, Anslinger received an honorary doctorate from Saint Francis College.

110. *New York Forum,* moderator Dallas Townsend, WCBS, April 28, 1962. AP, Box 1, File 8.

111. In the 1920s, the Prohibition Unit of the Internal Revenue Bureau included a Narcotic Division that was headed by Col. Levi G. Nutt, from 1920 until 1930. According to David Musto, "indiscretions" discovered in the operation of the Narcotic Division "led to Nutt's removal in 1930," the year that President Hoover approved creation of the FBN. David F. Musto, *The American Disease: Origins of Narcotic Control* (1973; reprint: New York: Oxford University Press, 1987), pp. 146-147, 207-208.

112. Garland Williams, interview with author, February 17, 1990.

113. Philip H. Glaessner Garcy to the Hon. Dr. Harry J. Anslinger, September 11, 1948. AP, Box 2, File 17.

Chapter 5

The FDA and the Practice of Pharmacy: Prescription Drug Regulation Before 1968

John P. Swann

Robert Fischelis, a leading spokesperson for professional pharmacy in the twentieth century, wondered in 1949 "whether there is anyone in the Food and Drug Administration [FDA] who knows anything about the practice of pharmacy or the practice of medicine or the physician-pharmacist-patient relationship with respect to prescriptions."[1] But in the following year, then Deputy Commissioner of Food and Drugs Charles Crawford remarked:

> The cry has been raised that our actions [of making an unauthorized prescription refill equal to an over-the-counter sale of a prescription drug] were disturbing the traditional physician-pharmacist-patient relationship. . . . In the cases about which we are concerned, there is merely a pharmacist-patient relationship, for the doctor plays no part in the transaction, while the pharmacist plays a double part: that of the physician, in determining that the patient needs additional medication, and that of the pharmacist, in supplying it.[2]

Versions of this chapter were presented at the symposium, The Prescription As a Focus of Historical Study, annual meeting of the American Institute of the History of Pharmacy, New Orleans, March 12, 1991, and at a seminar at the Francis C. Wood Institute for the History of Medicine, College of Physicians of Philadelphia, April 16, 1992. I am indebted to David Cowen, Wallace Janssen, Fred Lofsvold, Tom McGinnis, Ron Ottes, Robert Porter, Suzanne White, and James Harvey Young for their comments on this chapter. This chapter was published in *Pharmacy in History* 36 (1994): 55-70 and is reprinted here by permission from *Pharmacy in History,* a quarterly journal of the American Institute of the History of Pharmacy.

American Pharmaceutical Association (APhA) Secretary Fischelis and soon-to-be FDA Commissioner Crawford reflected a growing animosity between organized pharmacy and the FDA over an issue that cut to the heart of pharmacy practice: the regulation of prescription drugs. In the past, pharmacy organizations actually had worked closely with the FDA on several matters, such as the scope of the studies of the recently established drug laboratory of the FDA's predecessor, the Bureau of Chemistry of the Department of Agriculture, in 1903;[3] attempts to eliminate the so-called variation clause from the 1906 Food and Drugs Act; and efforts to address inconsistencies in the compounding of medicines by pharmacists. In fact, the very formation of the APhA stemmed in part from pharmacists' concerns about enforcement of the Drug Importation Act of 1848.[4]

With respect to refilling prescriptions for potent drugs, however, many pharmacists believed the FDA had unfairly excoriated professional pharmacy as a whole. Such a cold and bureaucratic federal agency seemed unable to appreciate or respect the prescription, the epitome of the doctor-pharmacist-patient relationship. On the other hand, from the FDA's standpoint, there were more than a few instances in which pharmacists destroyed lives and families, exceeding the bounds of ethics and law by refilling, sometimes incessantly, prescriptions for potent and dangerous drugs such as barbiturates, amphetamines, and sulfa drugs.

The issue of pharmacists refilling prescriptions without the authorization of the physician by no means began in the mid-twentieth century. In fact, to appreciate statements such as those of Fischelis and Crawford, we need to examine the context of prescription drug regulation and pharmacy practice, especially with respect to drug labelling. Prior to the 1840s, unauthorized refilling of prescriptions or sale without a prescription was unusual in the practice of pharmacy in America, "largely because apothecaries accepted their subserviant position to physicians, and the number of drugstores, although on the rise, had not yet reached the saturation point." Later, however, pharmacists' loyalty began to move from the physician toward the customer.[5]

PHARMACISTS, PHYSICIANS, AND EARLY PRESCRIBING AND DISPENSING PRACTICES

In 1867 the East River Medical Association of New York passed a resolution that requested pharmacists in that area to refrain from renewing prescriptions without the permission of physicians, or else association physicians would regard them as unworthy of patronage. The association circulated this resolution to all pharmacists in the district. By this time several Philadelphia physicians already were printing on their prescription forms the statement that the pharmacist could not refill the prescription without the written authorization of the prescriber. In this early postbellum period, other physicians petitioned for legislation at the state level to outlaw the indiscriminate and unauthorized refilling of prescriptions.[6]

Members of the American Pharmaceutical Association discussed the issue raised by the East River Medical Association at length. William Procter Jr. distilled many of the feelings expressed in the discussion:

> I can agree perfectly with these medical gentlemen, that much harm may come from the renewal of the prescription, but to pass any such stringent regulations as they have proposed in some of their meetings, making it penal on the apothecary to renew a prescription without the authority of the physician, I think is perfectly absurd in view of the present way of writing prescriptions, and the present amount of care that the physicians at large give to that branch of their duties.[7]

In the end, the APhA's official position on such efforts to restrict the pharmacist from refilling prescriptions without the prescriber's authorization recognized the problem while eschewing any formal way of dealing with it. The APhA promulgated the statement that it was

> neither practicable nor within the province of this Association [to endorse the East River group's plan]. Nevertheless we regard the indiscriminate renewal of prescriptions, especially when intended for the use of others than those for whom they were prescribed, as neither just for the physician nor the patient . . . such abuses should be discouraged by all proper means.[8]

At this time pharmacists preferred to deal with the problem of prescription refills by increasing educational and certification standards for pharmacy practice.[9] Formal academic training and the establishment of specific criteria for practice presumably would elevate pharmacists and weed out the greedy interlopers who were tainting the majority of honest pharmacists.

Such standards indeed were elevated in the late nineteenth and early twentieth centuries,[10] but the refilling of prescriptions remained an issue. Some argued that it was a question of ownership of the prescription,[11] and others voiced their opinions in terms of the harm this caused the business of a medical or even a pharmaceutical practice.[12] Physicians threatened to bypass pharmacists altogether and dispense their own medicines.[13] Pharmacists countered that the large number of dispensing physicians was in large part the reason why pharmacists refilled prescriptions on their own.[14] By the time of the Harrison Narcotics Act of 1914, several states already had laws prohibiting pharmacists from refilling prescriptions without the physician's permission, although these usually were confined to narcotics.[15]

The original code of ethics of the American Pharmaceutical Association in 1852 did not address the matter of refilling prescriptions, but the first major revision of the code in 1922 included several elements bearing on this issue. For example, according to the revised code pharmacists "should use every proper precaution to safeguard the public from poisons and from all habit-forming medications." More directly to the point, the pharmacist "should follow the physician's directions explicitly in the matter of refilling prescriptions. . . . Whenever there is doubt as to the interpretation of the physician's prescription or directions, he should invariably confer with the physician in order to avoid a possible mistake or an unpleasant situation."[16]

By the time of the revision of the code of ethics, selected drugs and prescriptions already had come under federal regulation. The 1906 act charged the Bureau of Chemistry with limited regulation of drug adulteration and misbranding, although how the bureau—and later the FDA—interpreted this charge changed over time. At first the bureau emphasized enforcement of food violations, principally because Harvey Wiley, chief chemist of the bureau, perceived these as greater dangers to the public health. This was despite the recent creation of the aforementioned drug laboratory within the bureau. Among drug activities, the agency typically focused on substandard synthetics and adulterated crude drugs and essential oils. The appointment of Walter Campbell in 1921 as acting chief chemist changed this philosophy. Under Campbell the bureau enforced the drug provisions of the 1906 act more aggressively, seeking manufacturers' compliance with the act through so-called contact committees between the bureau and trade associations.[17]

The Harrison Narcotics Act of 1914 brought federal regulation by an entirely different agency into the heart of pharmacy practice. Pharmacies wishing to dispense narcotics had to register with the director of the Internal Revenue Service and pay a nominal fee. A pharmacist could dispense a narcotic only with the written prescription of an authorized physician or dentist, although the law exempted certain preparations possessing less than a stated maximum amount of narcotic, as well as selected topical preparations. This was the first time such a stricture on prescriptions appeared in federal statutes.

The act prohibited orally issued prescriptions for narcotics, and it did not permit prescription refills. Pharmacy organizations objected to earlier versions of the act, although their concerns stemmed more from what they believed was its overall complexity than a preoccupation with the principle of an outsider controlling the narcotic prescription. In the end, pharmacy representatives supported the Harrison Narcotics Act.[18]

Prohibition also affected prescribing and dispensing habits of physicians and pharmacists. The 1919 Volstead Act and the Willis-Campbell Act of 1921 exempted medicinal alcohol prepared according to *United States Pharmacopoeia (USP)* and *National Formulary (NF)* standards, but implemented limits on the amount of wine and spirits that a physician could prescribe. Ordinarily a doctor could not write more than 100 prescriptions in ninety days, and a patient could not receive more than one pint of medicinal alcohol every ten days. Pharmacists, physicians, and wholesalers had to have permits to sell, prescribe, or dispense alcohol, and physicians were required to use special prescription blanks; the latter underwent a series of design changes in response to counterfeiting (see Photo 5.1).[19]

As mentioned earlier, organized pharmacy had an established record of cooperation with the Bureau of Chemistry and the FDA on drug regulation. For example, many pharmacists—and the American Pharmaceutical Association in particular—had long criticized the patent medicine industry. Some opponents admittedly attacked nos-

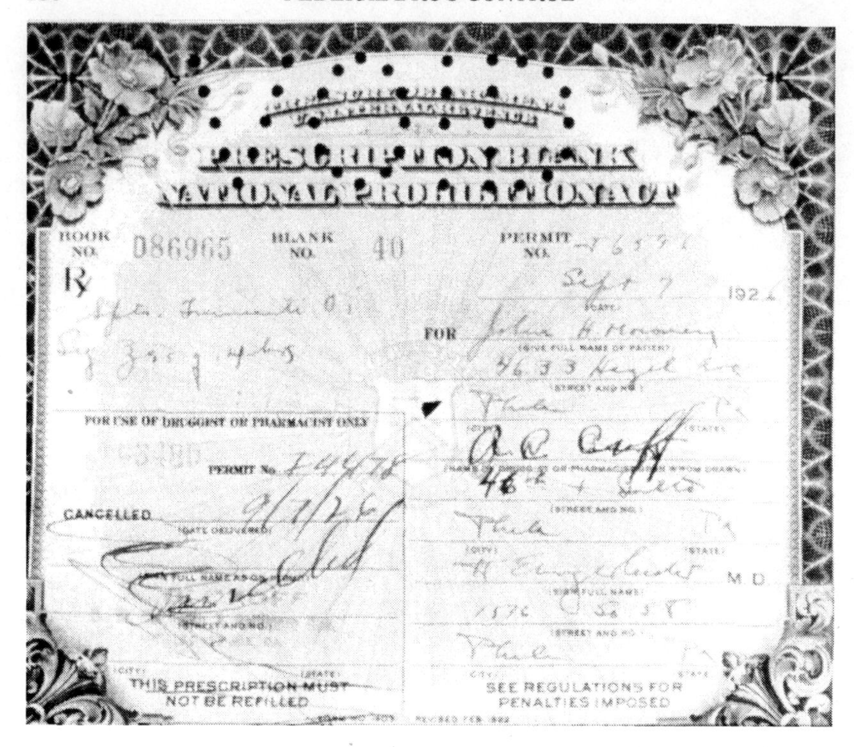

PRESCRIPTION BLANK
NATIONAL PROHIBITION ACT

BOOK NO. 086965 BLANK NO. 40 PERMIT NO.

℞

192

FOR

(GIVE FULL NAME OF PATIENT)

(STREET AND NO.)

FOR USE OF DRUGGIST OR PHARMACIST ONLY

PERMIT No.

CANCELLED

(DATE DELIVERED)

(STREET AND NO.)

(CITY) (STATE)

M.D.

(STREET AND NO.)

THIS PRESCRIPTION MUST NOT BE REFILLED

SEE REGULATIONS FOR PENALTIES IMPOSED

PHOTO 5.1. One version of the Internal Revenue prescription form, dated September 7, 1926; note the prohibition against refilling the prescription. (*Source:* Photo courtesy of the FDA History Office)

trums on the basis more of economics than morals: the proprietary medicines were siphoning sales from their prescription departments. But the APhA annual meetings nonetheless were a popular forum for members to decry the nostrums and their manufacturers. So, like many other groups, pharmacists played an important role in the coming of the 1906 act.[20]

One of the limitations of that act, a labeling provision known as the variation clause, led the APhA and the National Association of Retail Druggists (NARD) to collaborate with the bureau to try to eliminate this loophole. The variation clause enabled manufacturers to market drugs not in conformity with the standards of strength, quality, and purity according to the *USP* and the *NF,* as long as the variations from the standards were plainly stated on the label. These pharmacy organizations argued that the clause permitted companies to misrepresent their products as official to pharmacists and others, since manufacturers would print the official title in large type and then reduce the differences from the standards to fine print.[21] The campaign to eliminate the clause failed, and it remains in the law.

Another example of organized pharmacy's collaboration with the Bureau of Chemistry concerned the latter's routine investigation of pharmacy compounding practice. Attention to drug standards under the 1906 act on a retail rather than a manufacturing level was behind the bureau's inspections of the dispensing practices of pharmacists in the District of Columbia and Puerto Rico (which were under the bureau's jurisdiction). The bureau had made occasional investigations of these pharmacists since the 1910s to make sure the pharmaceuticals they were dispensing conformed to *USP* and *NF* standards. The agency typically reported on haphazard dispensing practices in these early surveys. Many of the more flagrant violations were sent to the courts. Hundreds of other cases that could have been adjudicated, according to the agency, instead were issued warnings by the bureau.[22]

Following yet another disappointing prescription survey in the District of Columbia in 1931, professional pharmacy worked with the FDA to correct these abuses. Inspectors delivered about a dozen different prescriptions for tinctures, elixirs, solutions, powders, and capsules to 100 pharmacies. The bureau, to investigate compliance with *USP* and *NF* standards, used these prescriptions, written by two District of Columbia physicians. The agency's survey revealed "great carelessness on the part of druggists in filling prescriptions." Sixty-seven of the 100 prescriptions were filled "unsatisfactorily, in one respect or another"; about half of these errors involved active ingredients that fell outside of *USP* tolerances for the prescriptions.[23] In response to these findings, the APhA formed a Committee on Prescription Tolerances in 1932 "to make a study of what constitutes reasonable deviations in prescription ingredients and to establish reasonable tolerances." The FDA cooperated with the APhA in this venture.[24]

However, the relationship between organized pharmacy and federal drug regulators deteriorated in the 1930s. Even prior to that, there were problems. For example, Prohibition witnessed illegal prescriptions and illegal sales of medicinal alcohol by physicians and pharmacists, as well as ersatz "drugstores" that became part of the liquor pipeline in America.[25] But the search for an amelioration of venereal diseases (VD) over the pharmacist's counter had a much more profound impact on the regulation of prescription drugs in America. Unauthorized refilling of prescriptions and over-the-counter sale of potent drugs for VD became a source of increasing concern for regulators, physicians, and pharmaceutical organizations by the 1930s.

Pharmacists often were the first health professionals consulted by victims of syphilis and gonorrhea. A 1933 survey by the American Social Hygiene Association of over 2,700 men across the country indicated that more men would first approach a pharmacist for treatment of VD than go to a physician or clinic.[26] Five years earlier, an investigation by public health and medical associations in the East Harlem district of New York revealed that one-fourth of all pharmacists treated VD patients.[27] Municipal and state pharmaceutical organizations decried such activities,[28] and the American Pharmaceutical Association had long voiced its disapproval by resolution and by its codified ethics: "The Pharmacist even when urgently requested to do so should always refuse to prescribe or attempt diagnoses."[29]

The therapeutic armamentarium for gonorrhea and a host of other bacterial infections changed drastically with the introduction of sulfa drugs in 1935. Following in the wake of the Elixir Sulfanilamide disaster,[30] the 1938 Food, Drug, and Cosmetic Act—and the regulations issued by the FDA to enforce this legislation—attempted to come to grips with sulfas and other therapeutic agents. Based on a concern for public health and the potential dangers of self-medication, the 1938 act deemed as misbranded any drug that (1) lacked adequate directions for use and (2) failed to carry adequate warnings of how its use could be unsafe to the patient. However, the act gave the FDA the authority to publish regulations to exempt any drug from the requirement of adequate directions for use.

Another provision mandated the statement, "Warning—May be habit forming," on drugs containing certain narcotic and hypnotic substances. In this case the act itself exempted any drug from this warning if it were dispensed on the written, nonrefillable prescription of a licensed physician, dentist, or veterinarian. Any drug that was dangerous when used according to instructions on the label was misbranded under the act. Finally, no new drug could be marketed without the approval of the FDA as to its safe use.[31]

PRESCRIPTIONS AND ADEQUATE DIRECTIONS FOR SAFE USE

The FDA issued regulations in December 1938 that specified how drugs could be exempted from the provision about adequate directions for use. The exempted product would have to include the warning, "Caution: To be used only by or on the prescription of a physician, dentist, or veterinarian." Also, the manner of labeling indications for the drug had to employ medical terminology that would not likely be understood by an ordinary person. By restricting access to the exempted drug through the prescription of a physician, dentist, or veterinarian, the FDA assumed that the prescriber would inform the patient of individualized directions for use.[32]

Even though these regulations mark the beginning of what we now recognize as the Rx legend, by themselves they did not mandate a new class of substances called prescription drugs (actually, the Harrison Narcotics Act accomplished that feat). Rather, the regulations informed manufacturers and distributors what had to be done if they chose to have their products exempted from the adequate labeling provisions of the act. The manufacturer had to put full and clear directions for use in labeling of any drugs to be used without medical supervision.[33] However, in a sense this was an academic question, because by the time these regulations appeared FDA already had taken steps toward restricting distribution of selected nonnarcotic drugs.

Those steps came in August and September of 1938, when the agency issued the first four of what eventually became hundreds of so-called trade correspondence, public policy statements issued to the industry, press, and whoever else might be interested in the issue. In this case, the agency informed distributors of sulfa drugs, aminopyrine and related products, and cinchophen and similar drugs that the sale of these dangerous pharmaceuticals for "indiscriminate" use by the general public would be subject to action under the 1938 act. Note that the APhA had also used the word *indiscriminate* in its condemnation of unauthorized prescription refills in 1868.[34] The agency argued that these drugs simply could not be labeled for safe use by the general public, and therefore would be made available only through medical supervision, i.e., by prescription. Manufacturers thereafter would have to label their sulfas, aminopyrine, and cinchophen with the Rx legend.

By 1941 FDA identified over twenty drugs or drug groups that were too dangerous to sell other than by a physician's prescription.[35] The FDA did not consider this so much a hard and fast list as a set of examples of drugs it deemed dangerous when distributed for self-medication. The policy of the agency was to refrain from compiling a set list of prescription drugs because "it would be scientifically untenable to attempt to draw up a fixed and final list of drugs since it must change as scientific facts are developed and further experience serves as a guide."[36] Such a policy was not exactly going to instill order. In any case, the FDA determined that certain drugs were "dangerous,"[37] basing its conclusions on "the consensus of qualified experts"[38] and "careful consideration."[39] Indeed, the agency appealed to physicians for their experiences, observations, and opinions to determine when a drug was dangerous.[40]

The FDA maintained that manufacturers should retain the responsibility for deciding whether to label a drug with the caution that it be dispensed only with a prescription.[41] After all, the firms had always been the ones to decide how a product would be distributed. In the past, manufacturers—or at least those in the ethical drug industry—had marketed (if not labeled) selected products such as insulin with the intention that they be used only under a physician's supervision.[42] In fact, as far back as the 1880s, a New York physician related his experience with detail men who assured him that their companies' preparations were sold only with prescriptions.[43]

But soon after the 1938 act, manufacturers began to abuse their responsibility for drug labeling. They were labeling many drugs that were safe for self-medication with the prescription legend. This could not have been all that surprising, considering how labeling requirements for self-medication had ballooned. For example, by 1940 the FDA issued twenty-eight detailed warning statements for over sixty laxatives, antitussives, analgesics, and other drugs and classes of drugs.[44]

Yet it was a mistake to believe (as many firms probably did) that one could avoid the detailed directions for use and the warnings by labeling pharmaceuticals with the Rx legend. Ordinarily, the FDA assumed that the prescriber issued directions and warnings orally to the patient. But if a prescription drug were dispensed with labeling that would permit the patient "to repurchase the article and continue to use it indefinitely and recommend it to friends" (e.g., without limiting refills of the prescription), the agency's philosophy was to expect the drug to be fully labeled for safe sale to the general public (i.e., with adequate directions for use and with all the appropriate warnings).[45] Manufacturers developed an even worse labeling practice, though: they were not labeling the same drug consistently. While one company might label a product with a prescription legend, another might indicate directions for use and warnings for self-medication on the same product.[46]

As debate over what eventually became the Durham-Humphrey Amendment increased in the late 1940s, the FDA lobbied in the face of considerable opposition from professional pharmacy, the American Medical Association, and the drug industry to have authority for compiling a list of drugs to be dispensed only under physicians' prescriptions. Among other reasons, the agency argued that the so-called administrative listing provision in the proposed law would simplify prosecution of illegal prescription drug sales by clearly identifying in a regulation those drugs restricted to prescription sales.[47]

The law and its enforcement thus appeared to abound in vagueness. The 1938 act was silent on the meaning of prescription drugs and refills, and many of the FDA's regulations were detailed yet at the same time only quasi-authoritative, the agency cautioning that these were "guides," "suggestions," or "examples." The Durham-Humphrey Amendment of 1951 helped correct this vagueness about prescription drugs and their refills. However, before discussing Durham-Humphrey, I consider how this confusing and disputed situation led professional pharmacy and the FDA to face off over the matter of illegal sale of prescription drugs and refills in the 1940s; in this case, with respect to a group of drugs named in the 1938 act—the barbiturates.

DANGEROUS DRUGS AND THE REGULATION OF BARBITURATES AND AMPHETAMINES

The first of the barbiturates, barbital (Veronal), was introduced into America in 1903 as a very useful hypnotic. Phenobarbital and numerous other barbiturates followed some years afterward, and this drug group soon was established as popular hypnotics, sedatives, anticonvulsants, and anesthetics. Despite the therapeutic versatility of these drugs, Goodman and Gilman's *Pharmacological Basis of Therapeutics* noted in 1941 "the ease with which the barbiturates can be prescribed by the physician and taken by the patient is probably their greatest disadvantage in that they are often employed when other sedatives or other therapy might be preferable."[48]

Industry's eagerness to modify the molecular structure of a proven drug and thereby gain a foothold in the market certainly was born out in the case of barbiturates. The basic barbiturate molecule was easily substituted with different side chains to produce mostly mildly different therapeutic effects. Thus, by 1947 firms had prepared over 1,500 different derivatives of barbituric acid; thirty of these were on the market, and seventeen barbiturates had received recognized status by the standard sources. Yet authorities in therapeutics advised that only a handful of these were really necessary.[49]

Evidence of chronic and acute barbiturate abuse began to appear in the American medical literature in the 1920s. Surveys of several major American hospitals during the periods 1928 to 1937 and 1940 to 1945 indicated that the frequency of acute barbiturate toxicity almost doubled by the later period.[50] By 1945 thirty-six states had laws controlling the distribution of barbiturates—all but one stipulating that a prescription was necessary for retail sale. Fifteen states either prohibited refills or permitted them only with the authorization of the prescriber. Only six laws addressed the distribution of barbiturates by wholesalers.[51] At the federal level, the 1938 act included the barbiturates in a list of drugs requiring the warning statement that they may be habit forming (see Photo 5.2). Within two years the FDA designated barbiturates as dangerous drugs.[52] That there was a problem was clear; the FDA dealt with the problem by focusing on abuses by pharmacists.

In the early and mid-1940s, the agency reported on occasional instances of pharmacists selling barbiturates and other dangerous drugs without prescriptions, and in 1946 the agency cited the first case of unauthorized refilling of a prescription for a barbiturate. The pharmacy in question repeatedly sold Nembutal capsules to a customer over a period of eighteen months on the basis of a single prescription.[53] A widely publicized case in 1945 from Waco, Texas, involved a pharmacy that had dispensed over 45,000 doses of barbiturates over an eighteen-month period without a prescription.[54] One of the more

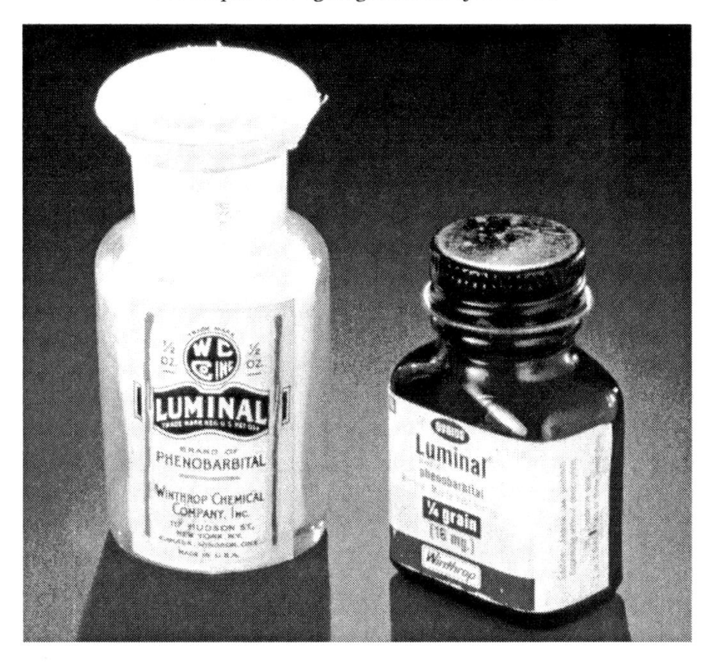

PHOTO 5.2. Two specimens of Winthrop Chemical Company's barbiturate, Luminal, from different eras are shown. The sample on the left dates from before the 1938 act; note its lack of any warning. The sample on the right bears the warning mandated by the 1938 act ("Warning: May be habit forming") and the revised proscription against unauthorized refilling under the 1951 Durham-Humphrey Amendment ("Caution: Federal law prohibits dispensing without a prescription"). (*Source:* Photo courtesy of the Division of Medical Sciences, National Museum of American History, Smithsonian Institution)

macabre reports told of a Kansas City woman who secured forty-three refills from two original prescriptions for barbiturates, one for fifteen capsules and another for ten capsules. Many of these refills were mailed to the woman from the same Los Angeles pharmacy where her original prescriptions had been filled before she moved. The drugs were so intoxicating that she was reduced to crawling, or so the height of the blood smears on her walls suggested. She was found dead in her home, partially eaten by rats.[55]

An FDA inspector typically conducted his investigation of a pharmacy with the intention of making the best case possible should a legal action result.[56] Thus the inspector usually made more than one buy, dressing and acting like an ordinary customer. The inspector made sure the pharmacist understood that the inspector was not sent by a physician; this way, there would be no question as to any prescriber involvement. After executing a number of buys, the "customer" would identify himself and begin an inspection of the pharmacy records.

The records search began by identifying the source of the drugs so as to establish interstate commerce (and thereby a federal case). Then, by inspecting the prescription files and the invoices for the prescription drug in question, it soon became clear how many sales could not be accounted for through legitimate (prescription) channels. Also, the FDA inspector tried to document the fact that the pharmacist should have realized that such sales were illegal. For example, any record of the pharmacist's membership in a professional pharmacy organization, or if the pharmacist otherwise subscribed to professional journals, would be incriminating evidence; both would have informed the pharmacist about the illegal sale of prescription drugs.[57]

The FDA's actions against pharmacies for illegal sales of prescription drugs theretofore had been comparatively restrained, since there was disagreement over the FDA's jurisdiction at this level of distribution. However, a ruling by the Supreme Court in 1948, the so-called Sullivan decision, clarified the authority of the FDA over retail sales of misbranded pharmaceuticals if the drugs had ever been in interstate commerce. In 1944, the venereal disease control officer at Fort Benning, Georgia, having discovered that soldiers were purchasing sulfa drugs directly from Columbus pharmacies for self-medication of gonorrhea, requested an FDA investigation into the matter. Posing as customers, FDA inspectors purchased sulfathiazole tablets from Sullivan's Pharmacy.

The district court convicted Jordan James Sullivan of violating the 1938 act, but the appellate court overturned this decision, claiming that the FDA had no jurisdiction at this level of commerce. The FDA appealed this decision to the Supreme Court, which ruled that retail outlets such as Sullivan's Pharmacy were indeed subject to the Food, Drug, and Cosmetic Act. This was necessary, the Court ruled, to protect the consumer.[58] From that point forward the FDA pursued illegal sales of barbiturates and other dangerous drugs in pharmacies with a new zeal.

According to the agency, inspectors encountered incredible tragedy over the illegal sale of prescription drugs such as barbiturates. Violations by pharmacists were causing "a sum total of more deaths, injuries, broken homes, human derelicts, and other tragedies than all the other violations of the [1938] act put together."[59] In 1950, for example, the FDA reported eighty-three prosecutions of illegal sales of prescription drugs. In part this increased activity was due to the leverage afforded by the Sullivan decision. But in part it was also the agency's response to professional pharmacy's threat to lobby for legislation that would remove jurisdiction over prescription practices from the FDA and give it to states.[60]

Robert Fischelis felt the FDA was treating professional pharmacy unfairly, that public statements such as the previous one tainted the whole profession. He did not disagree with the idea of banning the practice of unauthorized refilling of barbiturates and other dangerous drugs. Indeed, the APhA itself publicized in its journal the names of pharmacists who pedaled drugs over the counter illegally. Yet Fischelis argued that certain language, such as the FDA's statement about broken homes, unfairly cast aspersions on the vast majority of conscientious pharmacists.[61] In addition, he wanted the FDA to advocate that all drugs be sold under the professional supervision of pharmacists. The agency, Fischelis recalled, did not care where drugs were sold or by whom as long as they were properly labeled. He believed this philosophy compromised the public health.[62] Nothing in the law, of course, required the FDA to champion the monopoly of pharmacists over drug vending.

Such an argument perhaps was not all that surprising coming from the secretary of the APhA, but Fischelis's concerns indeed had some merit. The policy of when to investigate a pharmacy appeared to differ from the 1940s to the 1950s. By the 1950s, FDA inspectors typically moved against a pharmacy only if they had reports of possible problems at that establishment. Former inspector Fred Lofsvold observed that "most of the time, if I remember correctly, we only went to those drugstores where we had at least some kind of complaint. Preferably an injury, but at least some complaint, that they were illegally dispensing."[63]

However, during the 1940s enforcement was quite different; inspectors sometimes attempted to make purchases in broad sweeps of pharmacies in an area, whether or not they had reason to suspect those pharmacies of illegal sales of prescription drugs. A contemporary of Lofsvold and also an inspector at the time, Robert Porter, explained, "really much earlier [i.e., before Durham-Humphrey], the way we did it, we would just shop stores. We weren't building cases. It was an educational effort. . . ."[64] Fischelis and other defenders of professional pharmacy must have wondered if this approach were the most efficient use of the agency's limited resources, considering the FDA's own observation in its annual report for 1944 that "200 Federal inspectors are obviously unable to regulate the practices of over 50,000 retail drug stores."[65]

Some pharmacists clearly were guilty of selling barbiturates and other dangerous drugs illegally, but the distribution network responsible for the broken homes, injuries, and other tragedies was much more complicated than the public statements from the FDA conveyed. One link in the distribution network, the prescriber, received surprisingly little attention from the agency.[66] There was no statutory authority for the agency to pursue this venue in the drug distribution circuit. But some state public health officials and even other physicians blamed prescribers for much of the problem with the distribution of dangerous drugs like barbiturates.

For example, the use of "nonrepetatur" was not routinized in the prescription-writing habits of enough physicians.[67] Theodore Klumpp, chief of the FDA Drug Division, claimed that the FDA "encountered many instances of bromide intoxication and barbiturate habit-formation that were permitted to develop because physicians failed to place 'non-repetatur' on their prescriptions or otherwise clearly indicated how many times the prescription was to be refilled."[68] Also, the head of the State Board of Health and the chief of the Bureau of Food and Drugs in Indiana reported finding "that dentists and physicians have been lax in furnishing their patients with written prescriptions. Rather, the patient is told to go to the drug store and get a certain number of phenobarbital tablets."[69]

Bootleg sources of barbiturates—truck stops, cafés, bars, individual peddlers, and so on—also contributed to the problem. Klumpp suggested in 1945 that considerable quantities of barbiturates reached the public through this route. The FDA reported on many prosecutions of these distributors in the 1950s and 1960s, but little if any mention is made of actions against bootleg sources prior to Durham-Humphrey. Yet they certainly were around—and growing—throughout the 1940s.[70]

Investigations of illegal sales of dangerous drugs through nontraditional sources called for rather nontraditional methods, and FDA inspectors did what was necessary to fit in as believable customers. For example, beginning in the 1950s, selected inspectors attended truck driving school and specialized conferences to interdict illegal drugs (see Photos 5.3 and 5.4). Then-inspector Clifford Shane reflected on the experiences he and his cohort of faux truckers underwent in their unusual duties as food and drug inspectors:

> When I first came to work with FDA, the agency had sent a group to learn how to drive trucks so that they could go out on the highways and try and make buys of amphetamines at the truckstops. One of the problems that they encountered early on was that it became very apparent that they were not hauling a load and those trucks were empty. Anybody in a service station could see by looking at the springs that the trucks were empty, and they had a very difficult time making buys. What they did then was [load] the trucks with junk, and [they] also found out that the greatest thing was to sprinkle a little sugar on the tailgate. If anybody asked what you were hauling, you could always tell them sugar, because that would be contraband, especially down in the south. It was understood that you were probably going to make a delivery to a moonshiner. Buys were then very easy to make and we made a series of truckstop cases.[71]

CONCLUSION: LEGISLATIVE ANSWERS TO PRESCRIPTION DRUGS AND DANGEROUS DRUGS

In general, prescription drugs, prescription refills, and over-the-counter drugs were still a pharmaceutical muddle by 1951 even though (1) the concept of restricting drugs for sale by prescription only had existed in federal statutes since 1914; (2) a prohibition against refilling certain prescriptions had also existed since 1914; and (3) federal authorities mandated that more and more pharmaceuticals be labeled for sale by prescription only since 1938. The lack of definite statutes or court cases led the FDA to interpret what it understood to be a prescription, a prescription drug, and a prescription drug refill, but those interpretations did not satisfy pharmaceutical organizations.[72]

PHOTO 5.3. Though they may look like members of a Marlon Brando fan club, these are appropriately attired FDA inspectors attending a conference on undercover investigations of illegal distribution of dangerous drugs in Chicago, June 1951. Representatives from and former employees of the FBI, the trucking industry, and the FDA headquarters instructed the inspectors in criminal investigations and the problems of amphetamine use among truck drivers. (*Source:* Photo courtesy of the FDA History Office)

Resolving these differences created a significant rift between the FDA and professional pharmacy. The problem was not confined to the principle of the ethical, professionally trained pharmacist using his or her educated judgment to refill or refuse to refill physicians' prescriptions. There was also considerable debate over the definition of prescription drugs—who would define them, and how would they be defined? Moreover, pharmacy itself split over how best to resolve the disputes, NARD going one way and the APhA and much of the rest of organized pharmacy going another. This part of the story really warrants much more scrutiny than space permits here, but other sources explore this issue in detail.[73]

The Durham-Humphrey Amendment of 1951 did by statute what the FDA had begun to do by regulation thirteen years earlier by creating two categories of drugs.[74] The new law clarified the definition of a prescription drug: any habit-forming drug, any drug so toxic or harmful that it required the supervision of a practitioner for its administration, or any new drug approved under the safety provision of the 1938 act that had to be used under supervision. If at least one criterion

PHOTO 5.4. These FDA inspectors in undercover garb pose with their truck during the 1951 inspectors conference in Chicago. (*Source:* Photo courtesy of the FDA History Office)

applied to a drug, that product had to carry the statement, "Caution: Federal law prohibits dispensing without prescription." Refills of prescription drugs required the authorization of the prescriber.

Drugs that fell outside of the definition, that is, over-the-counter drugs, could not carry the prescription legend. All nonprescription drugs still required adequate directions for use and relevant warnings, but these could be sold directly to the consumer.[75] The initial decision to label a pharmaceutical for either over-the-counter sale or with the prescription legend still rested with the manufacturer—using the parameters cited in the new statute; however, the FDA issued advisory lists of prescription drugs. The courts would settle any unresolvable differences between the two sides. The FDA evaluated acute and chronic toxicity studies, age- and gender-specific therapeutic response, duration of use, and other clinical data to consider whether or not a product should be switched from prescription to nonprescription status. For example, the agency took these and other factors into account in its initial refusal to switch topical hydrocortisone from a prescription to an over-the-counter drug in the 1950s.[76]

Unauthorized refilling of prescriptions continued long after Durham-Humphrey, barbiturates and amphetamines accounting for the vast majority of drugs involved. The FDA's prosecutions of pharmacies for illegal sale of barbiturates, amphetamines, and other prescription drugs during the 1950s and early 1960s far exceeded prosecutions of all other sources (see Photos 5.5 and 5.6). One New York pharmacist

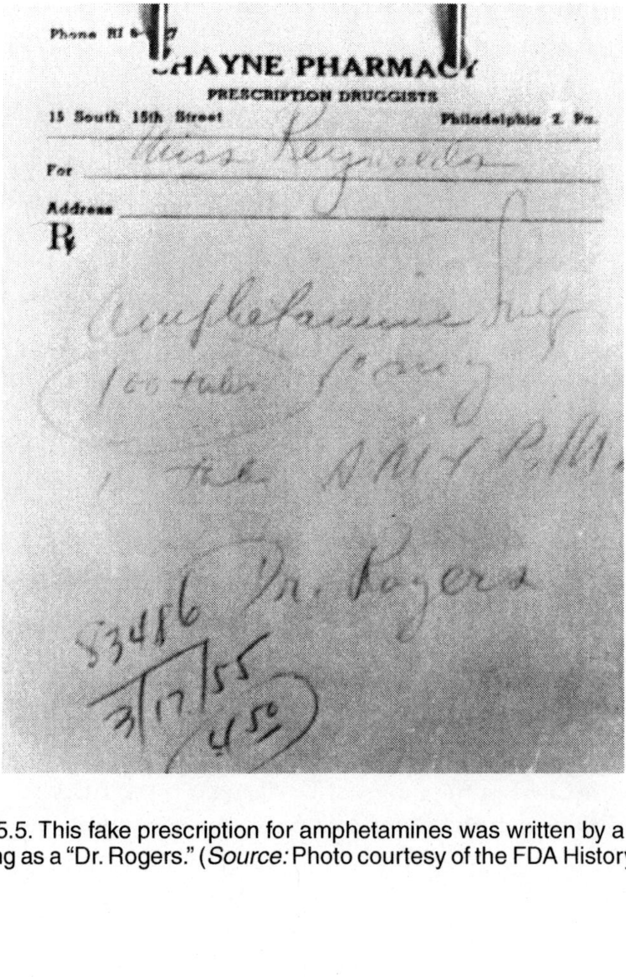

PHOTO 5.5. This fake prescription for amphetamines was written by a pharmacist posing as a "Dr. Rogers." (*Source:* Photo courtesy of the FDA History Office)

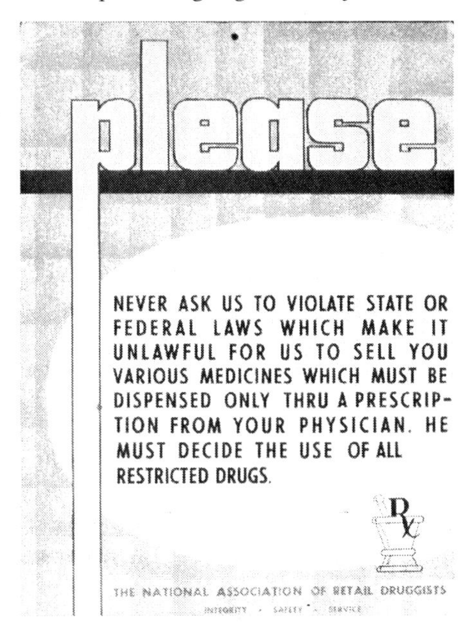

PHOTO 5.6. This is a placard from the National Association of Retail Druggists, circa 1950s, reminding the public of the physician-pharmacist-patient relationship. (*Source:* Photo courtesy of the FDA History Office)

claimed that over-the-counter sale of prescription drugs was not frequent but also not unusual in the several pharmacies where he worked in the 1960s.[77] Of course, it was much more difficult to pursue nontraditional sources such as truck stops, individual peddlers, bars, and so on. By the mid-1960s, however, the agency's actions against these nontraditional sources of dangerous drugs exceeded those against pharmacies.[78]

In 1965, influenced in part by the recommendations of the President's Advisory Commission on Narcotic and Drug Abuse two years earlier, Congress passed the Drug Abuse Control Amendments to the 1938 act.[79] This law attempted to control illegal distribution of amphetamines, barbiturates, and other nonnarcotic stimulants and depressants with a potential for abuse, as well as hallucinogens and counterfeit drugs. The FDA organized its enforcement efforts around a Bureau of Drug Abuse Control (BDAC), whose inspectors were given powers commensurate with the requirements of criminal investigation work. They could serve and execute search and arrest warrants, seize goods, and carry firearms. Theretofore, inspectors had depended on local and state law enforcement officers for these functions.

The BDAC lasted little more than two years in the FDA. In the first place, the FDA was responsible for countless other foods, drugs, cosmetics, and devices. The added burden of the supervision of licit and illicit traffic of so many drugs (literally hundreds eventually came under the 1965 amendments) alone was an awesome responsibility; the FDA did not fight to retain jurisdiction over dangerous drugs. Second, the separation of jurisdiction over narcotics and dangerous drugs was creating practical problems for investigators.[80] BDAC agents had authority over traffic in stimulants, depressants, and hallucinogens, but they did not regulate such drugs as heroin, marijuana, and morphine. Conversely, Bureau of Narcotics agents did not have jurisdiction over drugs covered under the Drug Abuse Control Amendments.

Unfortunately, drug dealers did not always operate as specialty vendors; some dealers might try to interest a BDAC agent in buying not only a hallucinogen but marijuana as well. Thus, unifying criminal investigations of drug trafficking under one roof was a pragmatic move. In April 1968, BDAC was transferred to the Department of Justice along with the Federal Bureau of Narcotics—which had existed in the Treasury Department since its establishment in 1930—to form the Bureau of Narcotics and Dangerous Drugs, the antecedent of the Drug Enforcement Administration.[81]

In conclusion, by the midpoint of the century two distinct classes of drugs emerged in American federal statutes. A large group of drugs thereafter would have to be dispensed through professional channels, but not all drugs as Fischelis had hoped.[82] Perhaps most surprising of all, little public attention was given to this new law at the time,[83] even though Durham-Humphrey effected fundamental changes in the practice of pharmacy and in the self-medication habits of every American.

URLS FOR SELECTED PRIMARY DOCUMENTS

Legal References on Drug Policy, Federal Court Decisions on Drugs by Decade, 1910
<http://www.druglibrary.org/schaffer/legal/legal1910.htm>

Legal References on Drug Policy, Federal Court Decisions on Drugs by Decade, 1920
<http://www.druglibrary.org/schaffer/legal/legal1920.htm>

Legal References on Drug Policy, Federal Court Decisions on Drugs by Decade, 1930
<http://www.druglibrary.org/schaffer/legal/legal1930.htm>

Legal References on Drug Policy, Federal Court Decisions on Drugs by Decade, 1940
<http://www.druglibrary.org/schaffer/legal/legal1940.htm>

Legal References on Drug Policy, Federal Court Decisions on Drugs by Decade, 1950
<http://www.druglibrary.org/schaffer/legal/legal1950.htm>

Pure Food and Drug Act of 1906
<http://coursesa.matrix.msu.edu/~hst203/documents/pure.html>

Harrison Narcotics Tax Act, 1914
<http://www.druglibrary.org/schaffer/history/e1910/harrisonact.htm>

NOTES

1. Robert Fischelis, "Straight from Headquarters," *Journal of the American Pharmaceutical Association: Practical Pharmacy Edition* 10 (1949): 209.
2. C. W. Crawford, "The Federal Drug Law and the Druggist," *Food Drug Cosmetic Law Journal* 5 (1950): 820.
3. Lyman F. Kebler, "Establishment of the Drug Laboratory in the Bureau of Chemistry, United States Department of Agriculture," *Journal of the American Pharmaceutical Association: Practical Pharmacy Edition* 29 (1940): 379-383. Reorganization of the Bureau of Chemistry in 1927 resulted in the transfer of regulatory authority to the newly formed Food, Drug, and Insecticide Administration. The name of the agency was shortened to the Food and Drug Administration in 1930.
4. J. H. Beal, "The American Pharmaceutical Association As a Factor in American Food and Drug Legislation," *Journal of the American Pharmaceutical Association* 26 (1937): 747-748. The 1848 law, passed before the establishment of the Department of Agriculture, was enforced by the Treasury Department.
5. Gregory J. Higby, *In Service to American Pharmacy: The Professional Life of William Procter, Jr.* (Tuscaloosa: University of Alabama Press, 1992), p. 14.

6. Editorial, *American Journal of Pharmacy* 39 (1867): 472-473; John K. Thum, "A Retrospect of Discussions on the Renewal of Prescriptions," *American Journal of Pharmacy* 79 (1907): 173-174; Minutes of the Fifteenth Annual Meeting of the APhA, New York, September 10-13, 1867, in *Proceedings of the American Pharmaceutical Association* 15 (1867): 112-114; and Minutes of the Sixteenth Annual Meeting of the APhA, Philadelphia, September 8-11, 1868, in *Proceedings of the American Pharmaceutical Association* 16 (1868): 66 (see the discussion by President Edward Parrish).

7. Minutes of the Sixteenth Annual Meeting of the APhA (1868), p. 64.

8. Ibid., p. 96. See also Editorial, *American Journal of Pharmacy* 40 (1868): 469-474; and Thum, "A Retrospect of Discussions," pp. 173-174.

9. For example, see Minutes of the Sixteenth Annual Meeting of the APhA (1868), p. 67.

10. See Glenn Sonnedecker, *Kremers and Urdang's History of Pharmacy,* Fourth Edition (Philadelphia: J. B. Lippincott, 1976), 226 ff., and Higby, *William Procter, Jr.*

11. W. F. Merrell, "Refilling Physicians' Prescriptions," *Massachusetts Medical Journal* 9 (1889): 392-394; Ward N. Choate, "The Pharmacist and the Law," *Bulletin of Pharmacy* 12 (1898): 392; William C. Alpers, "Prescription Repetition and Its Dangers," *Journal of the American Medical Association* 40 (1903): 1078-1079; G. A. Batchelor, "On the Ethics of a Prescription," *British Medical Journal* 2 (1905): 1697-1698; and "Ownership of Prescriptions," *JAMA* 109 (1937): 19B-21B.

12. Merrell, "Refilling Physicians' Prescriptions," p. 395; Thum, "A Retrospect of Discussions," p. 175; Batchelor, "Ethics of a Prescription," p. 1698; and J. F. Chandler, "Dispensing As an Art," *Journal of the Missouri State Medical Association* 30 (1933): 289. See also the resolution adopted by the East River Medical Association in 1867, stating that pharmacists were "injuring very materially the pecuniary interests of the profession, without gaining any particular benefit to themselves" (Editorial, *American Journal of Pharmacy* 39 [1867]: 473).

13. For example, see the primer on a physician's dispensing practice in Chandler, "Dispensing As an Art," pp. 288-291.

14. Charles H. LaWall, "What Professional Pharmacy Can Do for Medicine, and What It May Expect in Return," *International Clinics* 2, series 39 (1929): 244-252; and Charles H. LaWall, "Co-Operation Between Physician and Pharmacist," *American Journal of Pharmacy* 104 (1932): 48-53.

15. Alpers, "Prescription Repetition," p. 1079; L. F. Kebler, "Existing Laws Regulating the Sale of Habit-Forming Drugs and the Necessity for Additional Legislation," *American Journal of Pharmacy* 81 (1909): 189; Leslie Childs, "Liability of Physician for Selling Drugs Without a Prescription," *Journal of the Kansas Medical Society* 22 (1922): 10-11; and Lyman Kebler and Earl T. Ragan, *Drug Legislation in the United States,* U.S. Department of Agriculture, Bureau of Chemistry, Bulletin No. 98 (Washington, DC: Government Printing Office, 1906).

16. Code of Ethics of the American Pharmaceutical Association, adopted August 17, 1922, reproduced in the Challenge of Ethics in Pharmacy Practice, symposium presented at a joint session of the American Institute of the History of Pharmacy and the Academy of Pharmaceutical Practice of the American Pharmaceutical Association (Madison, WI: American Institute of the History of Pharmacy, 1985), p. 58. See also Batchelor, "Ethics of a Prescription," p. 1697. Charles H. LaWall, "The Ethics

of Pharmacy," *International Clinics* 1, series 33 (1923): 213-215, reproduces a code of ethics adopted in 1917 by medical and pharmaceutical associations in Australia that includes the following statement on the repetition of prescriptions: "When it is desired that a prescription should not be repeated, the prescriber should write on the prescription, 'Not to be Repeated,' or 'To be Repeated Twice Only,' or any specific number of times."

17. James Harvey Young, "Drugs and the 1906 Law," in John B. Blake (Ed.), *Safeguarding the Public: Historical Aspects of Medicinal Drug Control* (Baltimore: Johns Hopkins University, 1970), pp. 147-157.

18. David F. Musto, *The American Disease: Origins of Narcotic Control,* Revised Edition (New York: Oxford University Press, 1987), pp. 54-59, 121; James H. Beal, "The Common Law and Statutory Obligations of Pharmacists," in Frederick Peterson et al. (Eds.), *Legal Medicine and Toxicology by Many Specialists,* Second Edition, 2 volumes (Philadelphia: W. B. Saunders, 1923) volume 2, pp. 973-976; and Eric W. Martin and E. Fullerton Cook (Eds.), *Remington's Practice of Pharmacy,* Eleventh Edition (Easton, PA: Mack, 1956), pp. 1457-1460.

19. Andrew Sinclair, *Era of Excess: A Social History of the Prohibition Movement* (New York: Harper and Row, Harper Colophon Books, 1964), pp. 408-411; Bartlett C. Jones, "A Prohibition Problem: Liquor As Medicine, 1920-1933," *Journal of the History of Medicine* 18 (1963): 353-369; George Griffenhagen, "Medicinal Liquor in the United States," *Pharmacy in History* 29 (1987): 29-34; National Prohibition Act of 1919, Public Law 66-66 (41 U.S. Stat. 305), October 28, 1919; and An Act Supplement to the National Prohibition Act, Public Law 67-96 (42 U.S. Stat. 222), November 23, 1921.

20. James Harvey Young, *The Toadstool Millionaires: A Social History of Patent Medicines in America Before Federal Regulation* (Princeton, NJ: Princeton University Press, 1961), pp. 208-209. See also James Harvey Young, *American Self-Dosage Medicines: An Historical Perspective, Logan Clendening Lectures on the History and Philosophy of Medicine* (Lawrence, KS: Coronado Press, 1974), pp. 7, 17.

21. Glenn Sonnedecker, "Drug Standards Become Official," in *The Early Years of Federal Food and Drug Control* (Madison, WI: American Institute of the History of Pharmacy, 1982), pp. 28-39; Robert P. Fischelis, interview by James Harvey Young and Richard G. Hopkins, Ada, Ohio, September 17-19, 1968, transcript, National Library of Medicine, Bethesda, Maryland, pp. 28 ff.; and Food and Drugs Act of 1906, Public Law 59-384 (34 U.S. Stat. 768), June 30, 1906, Section 7.

22. Food and Drug Administration, *Federal Food, Drug, and Cosmetic Law: Administrative Reports, 1907-1949* (Chicago: Commerce Clearing House, 1951), pp. 324, 363, 381.

23. "Baltimore Station Checks Accuracy of Druggists," *Food and Drug Review* 16 (1932): 28. See also Young, "Drugs and the 1906 Law," p. 153. It is not clear what happened to the druggists who incorrectly dispensed the prescriptions, other than the fact that they were "cited to hearings."

24. "Prescription Tolerances," *Journal of the American Pharmacy Association* 22 (1933): 457-458; S. L. Hilton et al., "Report of the Committee on Tolerances," *Journal of the American Pharmacy Association* 22 (1933): 1051; and Samuel W. Goldstein, "The Professional Practice of Pharmacy," *American Journal of Pharmacy* 121 (1949): 174-175.

25. Sinclair, *Era of Excess,* p. 411.

26. F. J. Cullen, "Federal Control of Venereal Disease Nostrums Through Proposed Legislation," *Journal of Social Hygiene* 19 (1933): 520-521; see also George H. Bigelow and N. A. Nelson, "The Support of the Druggist in the Control of Gonorrhea and Syphilis," *New England Journal of Medicine* 203 (1930): 172; and "The Pharmacist and VD Control," *Journal of the Indiana State Medical Association* 35 (1942): 722.

27. Walter M. Brunet and Samuel M. Auerbach, "Unlicensed Practitioners and the Venereal Diseases," *Journal of Social Hygiene* 14 (1928): 104-119; see also E. W. Dittrich, "The Relation of the Physician and the Pharmacist," *Pharmaceutical Era* 45 (1912): 25.

28. Brunet and Auerbach, "Unlicensed Practitioners," pp. 105, 118-119; and Bigelow and Nelson, "Support of the Druggist," p. 172.

29. Code of Ethics of the American Pharmaceutical Association, August 17, 1922; see also Minutes of the Sixteenth Annual Meeting of the APhA (1868), p. 96.

30. James Harvey Young, "Sulfanilamide and Diethylene Glycol," in John Parascandola and James C. Whorton (Eds.), *Chemistry and Modern Society: Historical Essays in Honor of Aaron J. Ihde* (Washington, DC: American Chemical Society, 1983), pp. 105-125.

31. Federal Food, Drug, and Cosmetic Act, Public Law 75-717 (52 U.S. Stat. 1040), June 25, 1938, Sections 502, 503, and 505. For a comprehensive discussion of the evolution of this act, see Charles O. Jackson, *Food and Drug Legislation in the New Deal* (Princeton, NJ: Princeton University Press, 1970). With respect to drugs in particular, see Wallace F. Janssen, "Outline of the History of U. S. Drug Regulation and Labeling," *Food Drug Cosmetic Law Journal* 36 (1981): 428-430; and Peter Temin, *Taking Your Medicine: Drug Regulation in the United States* (Cambridge, MA: Harvard University Press, 1980), pp. 38-46, for substantially different perspectives on this legislation.

32. Theodore G. Klumpp, "The New Federal Food, Drug, and Cosmetic Act," *JAMA* 113 (1939): 2233-2234; Theodore G. Klumpp, "The Federal Food, Drug, and Cosmetic Act As It Applies to Drugs Dispensed by Physicians or on Physicians' Prescriptions," *JAMA* 116 (1941): 831; and 3 Fed. Reg. 3168, December 28, 1938.

33. This is in contrast to what Temin, *Taking Your Medicine,* pp. 46-47, claims. The regulations discussed not only how a drug could be exempted from the labeling provisions (which Temin quotes), but also what constituted adequate directions for use (which Temin does not quote); see 3 Fed. Reg. 3167-3168, December 28, 1938.

34. Trade Correspondence (hereafter TC)-1, August 26, 1938; TC-2, September 8, 1938; TC-3, September 8, 1938; and TC-4, n.d., all reproduced in Vincent A. Kleinfeld and Charles Wesley Dunn, *Federal Food, Drug, and Cosmetic Act: Judicial and Administrative Record, 1938-1949* (Chicago: Commerce Clearing House, c. 1949), pp. 561-562. On trade correspondence, see also Wallace Janssen, James Harvey Young, and Robert G. Porter, "Food and Drug Administration: Sources of Historical Information," 29 pp., typescript, 1985, files, FDA History Office.

35. TC-361, April 24, 1941, in Kleinfeld and Dunn, *Federal Food, Drug, and Cosmetic Act,* pp. 713-714. With respect to prescription drugs introduced after the 1938 Act, FDA would not approve any new drug intended for sale only on prescription if the labeling had any directions that might encourage unsupervised use; see Klumpp, "The Federal Food, Drug, and Cosmetic Act," p. 831, and Janssen, "History of U. S. Drug Regulation," p. 430.

36. TC-54, February 12, 1940, in Kleinfeld and Dunn, *Federal Food, Drug, and Cosmetic Act,* pp. 590-591; Theodore G. Klumpp, "The New Federal Food, Drug, and Cosmetic Act and the Practice of Medicine," *Ohio State Medical Journal* 37 (1941): 894 (quote); and Ruth de Forest Lamb, "What the Pharmacist Needs to Know About the New Food and Drug Law," *Modern Hospital* 54(1) (1940): 100.

37. TC-1.

38. TC-2.

39. TC-3.

40. Theodore [G.] Klumpp, "The Work of the Federal Food and Drug Administration," *Journal of the Medical Association of Georgia* 38 (1939): 279-280.

41. TC-361.

42. On how this concerned insulin, see Michael Bliss, *The Discovery of Insulin* (Chicago: University of Chicago Press, 1982), p. 173; and John P. Swann, *Academic Scientists and the Pharmaceutical Industry: Cooperative Research in Twentieth-Century America* (Baltimore: Johns Hopkins University Press, 1988), p. 139.

43. Henry C. Van Zandt, "Commercial Prescriptions," *Transactions of the New York State Medical Association* 2 (1885): 312.

44. Theodore G. Klumpp, "Significance of the New Federal Food, Drug, and Cosmetic Act," *Wisconsin Medical Journal* 39 (1940): 970; Chester I. Ulmer and Robert P. Fischelis, "How the New Food, Drug, and Cosmetic Acts Affect Physicians," *Journal of the Medical Society of New Jersey* 37 (1940): 413-414; TC-14, November 1, 1940, reproduced in Kleinfeld and Dunn, *Federal Food, Drug, and Cosmetic Act,* pp. 574-578; and Janssen, "History of U. S. Drug Regulation," pp. 432.

45. TC-330, September 5, 1940, reproduced in Kleinfeld and Dunn, *Federal Food, Drug, and Cosmetic Act,* pp. 699-700. FDA required warnings on the labels of certain prescription drugs; see American Medical Association Council on Pharmacy and Chemistry, "Prescription Writing," *JAMA* 140 (1949): 1157.

46. TC-361; "Information Concerning Drugs That Should Be Sold Only to or on the Prescription of Physicians, Dentists, or Veterinarians," *JAMA* 122 (1943): 324; Crawford, "Federal Drug Law," 812-813; and Richard Joseph Hopkins, "Medical Prescriptions and the Law: A Study of the Enactment of the Durham-Humphrey Amendment to the Federal Food, Drug, and Cosmetic Act," master's thesis, Emory University, Atlanta, GA, 1965, pp. 153-154.

47. Hopkins, "Medical Prescriptions and the Law," pp. 155-161.

48. Louis Goodman and Alfred Gilman, *The Pharmacological Basis of Therapeutics,* First Edition (New York: Macmillan, 1941), p. 146. See also J. W. Dundee and P. D. A. McIlroy, "The History of the Barbiturates," *Anaesthesia* 37 (1982): 726-734.

49. Samuel W. Goldstein, "Barbiturates: A Blessing and a Menace," *Journal of the American Pharmaceutical Association: Scientific Edition* 36 (1947): 5-6; Dundee and McIlroy, "History of the Barbiturates"; and Harold B. Gardner, "The Use and Abuse of Barbiturates," *Pennsylvania Medical Journal* 47 (1944): 451-452.

50. Charles O. Jackson, "Before the Drug Culture: Barbiturate/Amphetamine Abuse in American Society," *Clio Medica* 11 (1976): 56; and Goldstein, "Barbiturates," pp. 5-14.

51. Robert P. Fischelis, "A Review of the Present Status of Barbiturate Regulation," *Journal of the American Pharmaceutical Association: Scientific Edition* 35

(1946): 193-204. See also Lee D. Cady, "Barbiturates Should Be Sold Only on Prescription," *Journal of the Missouri State Medical Association* 36 (1939): 485-487, and "Regulation of the Sale of Barbiturates by Statute," *JAMA* 114 (1940): 2029-2036. California was the first state to pass a law controlling distribution of barbiturates, in 1929.

52. Federal Food, Drug, and Cosmetic Act, Section 502 (d), and TC-326, September 5, 1940, in Kleinfeld and Dunn, *Federal Food, Drug, and Cosmetic Act,* p. 698.

53. *Federal Food, Drug, and Cosmetic Law: Administrative Reports,* p. 1225. In 1943 the FDA brought its first case against a pharmacist for the illegal sale of a prescription drug under the 1938 act. This involved a Maine pharmacist charged with selling sulfathiazole tablets over the counter to sailors; see Hopkins, "Medical Prescriptions and the Law," pp. 6-7, and Douglas C. Hansen, interview by Robert G. Porter, September 26, 1978, transcript, National Library of Medicine, Bethesda, Maryland, pp. 53-56. On other examples of FDA's early investigations of pharmacists and illegal sales of barbiturates, see *Federal Food, Drug, and Cosmetic Law: Administrative Reports,* pp. 1003, 1059-1060, 1164, 1224.

54. Wallace Werble, "Waco Was a Barbiturate Hot Spot," *Hygeia* 23 (1945): 432-433; *Federal Food, Drug, and Cosmetic Law: Administrative Reports,* p. 1164; Walter Moses, interview by Ronald T. Ottes, May 21, 1987, transcript, National Library of Medicine, Bethesda, Maryland, pp. 22-24; and E. C. Boudreaux, "Termination of Prosecution Action," typescript, March 3, 1945, files, FDA History Office.

55. James Harvey Young, *The Medical Messiahs: A Social History of Health Quackery in Twentieth-Century America* (Princeton, NJ: Princeton University Press, 1967 [Expanded Edition, 1992]), pp. 272-273; and Crawford, "The Federal Drug Law and the Druggist," pp. 816-817. "Los Angeles Druggist Fined $2000," *Food Drug Review* 32 (1948): 230, claims the woman received ninety-three refills from this pharmacy. Fred Lofsvold, an inspector who investigated many cases of illegal sales of prescription drugs in the 1940s, believes the larger number of refills probably was more likely the case (personal communication to the author, July 29, 1991).

56. But note that the end result was more often than not a so-called PL-5 warning letter, at least prior to the Sullivan decision's clarification of the FDA's jurisdiction over retail pharmaceutical sales. See Clifford G. Shane, interview by Fred L. Lofsvold and Robert G. Porter, April 23, 1980, transcript, National Library of Medicine, Bethesda, Maryland, p. 7.

57. Robert G. Porter, interview by Fred L. Lofsvold, October 19, 1981, transcript, National Library of Medicine, Bethesda, Maryland, pp. 6-10. See also Fred L. Lofsvold, interview by Robert G. Porter, August 25, 1981, transcript, National Library of Medicine, Bethesda, Maryland, pp. 84-85.

58. Hopkins, "Medical Prescriptions and the Law," pp. 1-6; *Federal Food, Drug, and Cosmetic Law: Administrative Reports,* pp. 1233-1234, 1294, 1307-1308, 1335, 1358; and especially James Harvey Young, "Three Southern Food and Drug Cases," *Journal of Southern History* 49 (1983): 30-36. The Miller Amendment, Public Law 80-749 (62 U.S. Stat. 582), June 24, 1948, gave legislative strength to the Sullivan decision.

59. *Federal Food, Drug, and Cosmetic Law: Administrative Reports,* p. 1407; and *Food and Drug Administration, Annual Reports, 1950-1974, on the Administration of the Federal Food, Drug, and Cosmetic Act and Related Laws* (Rockville:

Food and Drug Administration, c. 1974), p. 14. See also Crawford, "The Federal Drug Law and the Druggist," pp. 815 ff.

60. Hopkins, "Medical Prescriptions and the Law," pp. 36-59 passim.

61. Fischelis, "Straight from Headquarters," p. 209.

62. Fischelis, interview by Young and Hopkins, pp. 65-66, 87-89.

63. Shane, interview by Lofsvold and Porter, p. 7.

64. Ibid. See also *Federal Food, Drug, and Cosmetic Law: Administrative Reports,* p. 1003; "California Pharmacy Board Convicts 12 Druggists," *Food and Drug Review* 28 (1944): 140 (a broad sweep by state inspectors); and "Codeine-Phenobarbital Kills 9-Year-Old Boy," *Food and Drug Review* 29 (1945): 93.

65. *Federal Food, Drug, and Cosmetic Law: Administrative Reports,* p. 1106.

66. Cf. "Barbiturates Result in Death," *Food and Drug Review* 32 (1948): 314, which relates the story of two young lovers who obtained prescriptions for barbiturates from each of five different physicians and then attempted to commit suicide with the drugs. After reporting the death of the man and near-death of the woman, this notice in the FDA newsletter concludes with the curious statement, "No refills were involved."

67. Howard T. Phillips, "A Doctor Looks at Counter Prescribing," *West Virginia Medical Journal* 28 (1932): 75; W. Henry Rivard, "If I Were a Medical Man," *Rhode Island Medical Journal* 28 (1935): 2; E. Fullerton Cook, "The Importance and Advantages of Prescription Writing in Medical Practice," *JAMA* 107 (1936): 967; and Goldstein, "Barbiturates," p. 7.

68. Klumpp, "New Federal Food, Drug, and Cosmetic Act and Practice of Medicine," p. 894.

69. John W. Ferree and J. C. Schneider, "The Part of Physicians and Dentists in the Dangerous Drug Program," *Journal of the Indiana State Medical Association* 34 (1941): 695; and G. Wilse Robinson, "Observations on Addiction to Barbituric Acid Derivatives," *Journal of the Missouri State Medical Association* 36 (1939): 492-493.

70. Jackson, "Before the Drug Culture," pp. 50-52; "APhA Holds Conference on Barbiturates," *Journal of the American Pharmaceutical Association: Practical Pharmacy Edition* 6 (1945): 358; and Young, *Medical Messiahs,* pp. 279-280.

71. Shane, interview by Lofsvold and Porter, p. 6. See also Hansen, interview by Porter, pp. 56-70.

72. Paul B. Dunbar, "What Does the Federal Law Require on Labeling and Refilling," *American Druggist* 119(3) (1949): 78-79, 134, 136, 138, 140; Hopkins, "Medical Prescriptions and the Law," pp. 60-102; and 15 Fed. Reg. 8651-8652, December 6, 1950.

73. See Hopkins, "Medical Prescriptions and the Law," and Fischelis, interview by Young and Hopkins.

74. An Act to Amend the Federal Food, Drug, and Cosmetic Act, Public Law 82-215 (65 U.S. Stat. 648), October 26, 1951.

75. Ibid.; 17 Fed. Reg. 1130-1133, February 5, 1952; and 17 Fed. Reg. 6818-6820, July 25, 1952.

76. Young, *Medical Messiahs*, pp. 277-278; and Charles Wesley Dunn, "The New Prescription Drug Law," *Food Drug Cosmetic Law Journal* 6 (1951): 967. Cf. Peter Barton Hutt, "A Legal Framework for Future Decisions on Transferring Drugs from Prescription to Nonprescription Status," in *Rx OTC: New Resources in Self-*

Medication . . . A Symposium, 1 November 1982, Condensation of Papers and Discussion (Washington, DC: Proprietary Association, c. 1982), pp. 34-35, who argues that the FDA evaluated the prescription or nonprescription status of a pharmaceutical mostly on an ad hoc basis, without elaborating any systematic rules for these decisions other than what the 1951 amendment stated.

77. Personal communication to the author.

78. E. Barrett Prettyman et al., *The President's Advisory Commission on Narcotic and Drug Abuse: Final Report, November 1963* (Washington, DC: Government Printing Office, c. 1963), pp. 44-45; and FDA, Annual Reports, 1950-1974, pp. 261-542 passim. For the period 1960-1965, the agency's prosecutions against nontraditional sources represented 49 percent of the total, while actions against pharmacies were 47 percent and those against physicians 4 percent.

79. E. Barrett Prettyman et al., *The President's Advisory Commission on Narcotic and Drug Abuse;* Drug Abuse Control Amendments of 1965, Public Law 89-74 (79 U.S. Stat. 226), July 15, 1965.

80. Edward Wilkens, personal communication, March 29, 1990.

81. The discussion of drug regulation under BDAC is drawn from John P. Swann, "The Regulation of Barbiturates in the U. S., 1938-1968," presented at the Pharmacy World Congress, Washington, DC, September 4, 1991.

82. Fischelis, interview by Young and Hopkins, pp. 87 ff.

83. Janssen, "History of U. S. Drug Regulation," p. 435.

Chapter 6

Habitual Problems: The United States and International Drug Control

William B. McAllister

INTRODUCTION

The international history of drug regulation traces back over a century and can profitably be divided into four general periods. Between roughly 1880 and 1920, governments and nongovernmental actors strove to define the nature of the drug problem and craft initial policies to deal with the perceived negative effects of abuse. From around 1920 to the mid-1930s, states and international organizations negotiated an international regulatory regime that supervened and channeled national legislation. During the remainder of the 1930s through the mid-1960s, governments and other actors attempted to implement these international/national rules, continually modifying their content and application in light of changing circumstances. Beginning in the later 1960s, a global eruption of drug use caused the system's participants to reconsider their aims and practices. In one sense the century closed much as it began, with differing visions about and solutions for the international drug problem competing for attention. Yet the nature and scope of the drug question changed radically during the twentieth century. Governments and international agencies constructed massive bureaucracies, engaged in considerable legislative activity, and attempted to implement policies intended to change the behavior of millions of individuals, with varying degrees of success.

Three key themes appear in this chapter. First, U.S. lawmakers and administrators were acutely aware of the interrelation between their domestic initiatives and global trends. American activities can be best understood by taking into account the international context within which contemporaries operated. Second, nonmedical considerations often played a key role in determining the nature and application of drug-related legislation. A variety of political, economic, sociocultural, strategic, religious, and other factors impinged on drug policymaking at both the international and domestic levels. Finally, U.S. attitudes and actions have usually not been unique, but have typically reflected larger world trends. Although the United States often played the role of leader (and sometimes laggard), American drug-control attitudes and activities largely paralleled those elsewhere, especially in Western industrialized countries.

U.S.–world interaction concerning drug regulation represents the typical "attention pattern" replicated elsewhere, especially in the industrialized West. In the late nineteenth-early twentieth centuries American authorities attempted to devise local/national solutions, only to discover that the nature of the drug problem made it necessary to engage in international negotiations. From 1900 through 1920, American control advocates accomplished some of their goals through multilateral talks, joint efforts, and treaty negotiations, but could not achieve their more far-reaching policy objectives that attempted to solve the domestic "drug problem" by eliminating excess supplies overseas. That disappointment led legislators and bureaucrats to redouble their efforts to ensure domestic action on key policy initiatives between the mid-1910s and the early 1920s. Again feeling the need for international support, the United States engaged in high-profile activity at the League of Nations from 1923 to 1925. Frustrated in their hopes for thoroughgoing international reform, American policymakers turned to domestic measures until 1930-1931. At that time a combination of bureaucratic, economic, and strategic imperatives impelled greater U.S. participation in transnational negotiations. This pattern recurred, with major domestic activity taking place during the later 1930s and early 1940s, the mid-1950s, and from the later 1960s through the early 1970s. Important international initiatives took place during the early 1940s through the early 1950s, during the mid-1960s, and in the early 1970s. Since that time, the progression has been more mixed, but one can nevertheless discern the yin-yang nature of drug policy: success or failure on the domestic front leads to new activities in the international arena, and vice versa. The "inward" and "outward" periods do not correspond precisely, but the general pattern recurs. Most governments have gone through similar undulations as they attempt to wrestle with the complexities of the drug question.

1880 TO 1920: DEFINING THE ISSUES

The development of pharmaceutical industry capabilities in the United States in the late nineteenth century, described by Spillane, mirrored that taking place overseas. In Europe, German, Swiss, British, French, and Dutch firms competed for market share on a national basis, invested in research, enhanced manufacturing procedures, devised new drug-delivery systems, and advertised aggressively. A nascent Japanese pharmaceutical sector evolved along similar lines.[1]

Moreover, European and Japanese companies expanded their operations abroad. Especially in the case of coca, pharmaceutical firms competed for access to raw materials, often developing new agricultural production capacity in the process. Companies developed and exported preprocessing techniques that enabled them to concoct semifinished extracts overseas for shipment to home-country factories. The industry promoted its wares with aplomb, both within the rising profession of medical practitioners and among the general populace. Raw materials such as opium and coca were treated as commodities traded on the international market.

Thus, by the early twentieth century a substantial global pharmaceutical drug trade existed, dominated by some of the world's original multinational corporations. Led primarily by European-based pharmaceutical firms including Hoffmann-LaRoche, Merck, Bayer, T. Whiffen, C. H. Boehringer, and Nederlandsche Cocaine Fabriek, companies competed for business on the basis of price, availability, and quality of product in a worldwide market. American firms' participation in the international trade remained limited, but U.S. companies such as Mallinkrodt Chemical Works, New York Quinine, Merck (opiates), and Maywood Chemical (coca products) were acutely aware of both the possibilities for enhancing their business through exports and the threat posed by imports from overseas.[2]

The success of the pharmaceutical companies in selling their wares abroad made the potential for widespread drug abuse more apparent. For example, China took serious measures to curb opium smoking after 1907, but many users switched to heroin or morphine. In the absence of any national or international restrictions, pharmaceutical firms manufactured and sold as much of those drugs as the market would bear. Many Westerners feared that an unfettered trade in addicting substances would cause similar increases in drug abuse at home. American efforts to impose domestic controls took place within this international context that featured, in China, a compelling example of the potential negative consequences of inaction.[3]

The trend toward drug regulation in the United States paralleled that in other countries. By the late nineteenth century, control of drug distribution and use became a pressing concern in Canada, many European countries, Japan, China, and among colonial administrations. The limitations of local enforcement became clear; stringent regulatory activities foundered because neighboring lax jurisdictions served as reservoirs of both supply and demand. Consequently, a general consensus favoring uniform national or colonywide control measures emerged. Policymakers in many countries encountered the same difficulties that beset their American counterparts: they attempted to define legitimate and illegitimate use, questioned the medical efficacy of addictive substances, witnessed social prejudice directed toward user groups, expressed fears about recreational drug use, encountered rising professionalization among physicians, pharmacists, and government bureaucrats, and confronted the influence of pharmaceutical industries. Most national and colonial governments enacted drug legislation reforms between 1890 and 1920. Those measures attempted to achieve the same effect as the United States sought to accomplish with the 1906 Pure Food and Drug Act and the 1914 Harrison Narcotics Act by safeguarding food supplies, ensuring medicinal quality, and regulating access to addicting substances. In some countries, the medical profession retained more authority to determine how laws would be interpreted and enforced, while in others government agencies occupied the key sites of control. In any case, by the early twentieth century the global trend was clear: the question was not whether access to drugs ought to be regulated, but what level and type of regulation was appropriate.[4]

Yet the initial impulse toward international control owed more to geopolitical and commercial factors than to medical concerns. The international political and economic problems caused by opium abuse in China acted as the key catalyst in the formation of the international drug control regime. Examining how the "Great Game" in East Asia related to drug control issues illustrates the myriad factors that drug policymakers from all countries, including the United States, had to consider.

By the late nineteenth century, administrative impotence, internal disruption, and external pressure caused increasing governmental debility in China. Unfettered imports and sale of opium contributed significantly to Chinese weakness. As central authorities became progressively more feeble, it appeared that the country might disintegrate. Imperialist governments with a limited capacity to project power, including Russia, Japan, France, and Germany, did not find the prospect of a divided China unappealing. Each hoped to carve out spheres of influence that would benefit domestic constituencies.

Two important states, however, wished to see Chinese territorial integrity remain intact. The strongest nation of the era, Great Britain, surpassed all its rivals in sales and investments throughout the Middle Kingdom. A breakup of China would harm British prospects for widespread influence. The United States, possessing no ability to project power onto the Asian mainland, nevertheless hoped to gain access to Chinese markets. American officials supported antiopium efforts in hopes of currying favor and trade advantages with Chinese officials. Washington and London championed early international drug control efforts precisely because curbing the opium problem complemented their national economic-strategic interests in East Asia.

In addition to such geopolitical concerns, calls for international control came from other quarters. Western missionaries and their supporters, which included many medical missionaries, decried opium's effect upon the populace. They denounced Western complicity in the drug trade as immoral and blamed rampant opium abuse as a chief reason for their unsuccessful efforts to convert more Chinese to Christianity. In the Netherlands and Great Britain, religious missionary groups successfully pressured governments to reform colonial drug administrations. Supporters of the Liberal party in England advocated cessation of Indian opium exports, on both religious and moral grounds. Trading houses that did not traffic in drugs also supported controls in the hope that a more sober Chinese populace would buy other foreign goods in lieu of opium.

As a result of all those factors, but primarily because of geopolitical and commercial calculations surrounding the Great Game in East Asia, the leading Western powers lurched toward an agreement concerning drug control. The Americans led the way, with Great Britain playing a less enthusiastic second. At the Hague during the winter of 1911-1912, the principal colonial powers and China agreed to the first treaty designed to control the international trade in addicting substances.

Material interests, however, played a larger role than medical considerations among those negotiating the treaty. The British did not want to suffer international opprobrium for opposing the curtailment of opium shipments from India to China, so they expanded the agenda by insisting that manufactured drugs also be included in control legislation. London calculated correctly that Berlin would object because Germany dominated the international trade in pharmaceutical opiates such as morphine, heroin, and codeine. In order to draw resistance from the Dutch and the South American states, the Germans insisted that cocaine controls be included. Ultimately such amendments weakened the treaty. In order to get all key agricultural producer and pharmaceutical manufacturer nations to sign, the resulting agreement was couched in vague terms. Various provisions called upon signatories to license manufacturers, regulate distribution, and halt exports to those jurisdictions that prohibited imports, yet the treaty did not create a uniform system to accomplish those goals. Governments retained wide latitude to interpret key clauses as they saw fit; the treaty required states merely "to make their best efforts" to curb the trade and set no timetable for completion of that task. German protests ensured that codeine escaped regulation altogether, and governments representing major agricultural producing areas such as Great Britain, Turkey, and Persia (now Iran) defeated language that required limitation of cultivation. The 1912 Hague Opium Convention, typically for a pact breaking new ground, contained many loopholes for those not wishing to institute meaningful controls.

Despite the shortcomings of the Hague Conference and other multilateral negotiations, American control advocates played the "international card" to great advantage in their quest to secure domestic legislation. Between 1909 and 1914, Dr. Hamilton Wright, a physician who adopted the drug issue as his special cause, spearheaded efforts to create a law that would pass constitutional challenges. Prevailing interpretations of the U.S. Constitution restricted the exercise of federal authority; no justification for national police enforcement existed at that time. Wright and his allies repeatedly noted that the State Department could hardly press other nations to impose more stringent drug control regulations while the United States had not enacted a federal law. The issue became acute after 1912 when the

American delegation played a leading role in the Hague negotiations and signed the treaty in anticipation of Senate ratification. Wright's invocation of international obligations spurred disputing parties to arrive at a compromise. The 1914 Harrison Narcotics Act, which provided the functional basis for federal regulation, was passed in large measure as a response to calls for America to comply with its international responsibilities and as an example for other governments to emulate.[5]

The effects of World War I contributed to the perceived need for substantive national and international drug control measures. Germany produced the bulk of the world's opiate and coca-based products until 1914, but the wartime blockade cut off supplies of vital medicines from that quarter. During the war, the British, French, Japanese, Dutch, Swiss, and Americans increased their manufacturing capacity and enhanced their research and development programs. Many feared that when the war ended these significantly augmented manufacturing and research capabilities would flood the market with drugs, leading to an increase in abuse. Many governments enacted or enhanced their domestic drug control legislation in the name of prosecuting the war effort. Great Britain utilized the 1914 Defense of the Realm Act to impose restrictions on the sale and export of addicting substances in 1916. Canada used the 1914 War Measures Act to curb excess imports in 1919. The German Imperial Health Office instituted regulations to conserve the supply of vital opiates, including measures to curb nonmedical use injurious to the war effort. The League of Nations, created as part of the peace process, would become the venue in which international negotiations concerning drug control took place in the 1920s.[6]

Consequently, by the end of World War I the interplay between international and national considerations had already proven crucial to the development of drug control efforts, and the key factors animating drug control had become apparent. In the United States, the 1912 Hague Opium Convention assisted pro-control advocates, providing a key rationale for passage of the 1914 Harrison Narcotics Act. In subsequent years the U.S. government would often claim treaty obligation as a justification for enacting federal drug control legislation. American drug control advocates also attempted to use international treaty negotiations as a vehicle for imposing U.S.-style drug policy on other governments. Yet the drug question could not be separated from its global political, economic, technological, sociocultural, religious, and strategic contexts. The complications resulting from the many nonmedical considerations that affected drug control proved crucial to the outcome of events both domestically and abroad.

EARLY 1920S TO MID-1930S: SHAPING THE INTERNATIONAL CONTROL REGIME

By the early 1920s, American officials, like most contemporaries, concluded that enacting effective drug control required a truly international effort. If some governments imposed restrictions but others did not, the traffic would simply gravitate to those jurisdictions featuring less restrictive legislation.[7] In a manner reminiscent of domestic developments in preceding decades, states sought to define the parameters of legitimate commerce in addicting substances. Between 1920 and 1931, governments negotiated a set of international rules for the licit drug trade, thereby also defining what constituted the illicit traffic and providing an overarching structure for national control efforts. Yet attempts to forge an international consensus on drug control did not occur in a vacuum. During the interwar years American officials, like their counterparts abroad, could not escape the political aspects entwined in the drug issue.

A principal area of controversy concerned the League of Nations itself. After World War I, the league served as a key venue for negotiations about a variety of transnational topics, including the drug question. Yet the U.S. government rejected the Versailles peace treaty that included the League Covenant and refused to join the league. Relations with the Geneva-based organization remained strained throughout the 1920s and 1930s. The American desire to play a leading role in international drug control efforts clashed with widespread domestic antileague sentiment.

The American love-hate relationship with the league affected domestic and international drug policy. During the early 1920s, the U.S. government refused to cooperate with league efforts to enact arms control, to stabilize the international economy, and to promote peaceful resolution of disputes. By 1923, however, American policymakers reconsidered their position, owing to rising tension between France and Germany over implementation of the Versailles peace treaty, increased turmoil in East Asia, and the failure of several independent

American initiatives to foster international cooperation. The Harding administration approved formal American participation in league drug-control activities as one of several exploratory interactions with Geneva.

During a series of negotiations lasting from 1923 through early 1925, U.S. delegates, led by House Foreign Affairs Committee chairman Stephen Porter, adopted an extreme supply-control position in an effort to limit the distribution of illicit drugs. The Americans advocated a drastic reduction in agricultural production, calling for worldwide elimination of all opium and coca cultivation in excess of that necessary for medicinal purposes. U.S. representatives also wanted to end opium smoking, despite its pervasive use as a substitute for more sophisticated medicinal treatment in many parts of Asia. Majority opinion in the American medical community had already concluded that the prognosis for rehabilitating addicts appeared bleak, so it stood to reason that all drugs not required for strictly therapeutic purposes should be kept out of the country. Americans claimed the best way to do that was to eradicate all nonmedicinal supplies and usage as defined by Western criteria.

In the early 1920s the League had initiated investigations into the nature of addiction that might have developed alternative etiological conceptions, but American intervention polarized the debate. Many colonial officials recognized that an extreme supply-control position was impossible to implement because they could not impose substantive restrictions in remote backcountry areas. Moreover, implementing radical supply control measures would threaten the interests of agricultural producers in Asia and Latin America, hurt traders who bought and sold addicting substances, impoverish governmental authorities who profited from the traffic, and likely reduce the profits of pharmaceutical companies. In the rush to protect economic, political, and other interests, all sides eschewed etiological investigations. They concentrated instead on the supply-control paradigm, adopting either pro- or anticontrol positions.

Representatives from an impressive array of forty-one governments met in Geneva during the winter of 1924-1925 to negotiate a new international drug control treaty. The Americans hoped the agreement would supersede the weak provisions of the 1912 Hague Opium Convention. They belligerently advocated eliminating opium smoking and instituting rigorous control over agricultural production, but delegates from most other governments rejected those proposals as impossible to implement. In exasperation, the U.S. delegation withdrew from the conference, causing a long-lasting rift between Washington and other governments fundamentally in favor of international cooperation on drug control but opposed to the radical American agenda.

The remaining delegations fashioned an important treaty, the 1925 International Opium Convention. The accord established a standardized import-export certification system designed to regulate drug movements; all signatories had to compile statistics on drug transactions passing across their borders and keep records of stocks within their countries according to a uniform procedure. The treaty created the Permanent Central Opium Board, an international body of experts that examined those import, export, and inventory statistics in order to oversee the licit trade and curb illicit trafficking. The 1925 convention also established procedures enabling governments to add additional drugs to the list of controlled substances so that further treaty negotiations would not be required for every new addicting drug.[8]

For the remainder of the 1920s, relations between the United States and those governments that supported League drug-control efforts remained strained. Although never ratifying the 1925 treaty, the United States complied with many of its provisions. Nevertheless, American officials remained disappointed with what they considered tepid international cooperation, and consequently they turned to domestic solutions to curb the domestic drug problem. Porter's initiatives, discussed in Chapter 2, to create the federal narcotics farms and the Federal Bureau of Narcotics (FBN) represent, in part, a frustration with the inability of American diplomacy to enact its supply-control agenda abroad.

Upon Porter's death and with the creation of the FBN, both of which occurred in the summer of 1930, countervailing domestic and overseas influences altered somewhat the American attitude toward international cooperation. Harry Anslinger, commissioner of the FBN, sought to cultivate widespread domestic support for his agency. On the one hand, he espoused many traditional "hard line" policies. He treated the drug problem strictly as an enforcement issue, concentrating his efforts on policing the licit trade, prosecuting illicit traffickers, and marginalizing drug abusers. Anslinger eschewed any serious discussion about alternative approaches such as medicalization, treatment, or prevention. The commissioner solidified a longstanding American propensity to "export" responsibility for the U.S. drug problem by blaming excess agricultural production overseas. It should be noted that most Americans supported Anslinger's policies and that such attitudes were commonly held in Canada, Japan, and many Western European states.

But Anslinger also forged close ties with U.S. pharmaceutical firms that ultimately compromised his own supply-control prescriptions. He secured pharmaceutical firms' cooperation in enforcing control provisions in exchange for promoting increased American exports abroad. Anslinger also supported pharmaceutical companies' desire for reasonably priced raw materials, primarily opium and coca. Ensuring price competition among suppliers, however, required excess capacity among agricultural producers. To deal with these contradictions, Anslinger advocated a combination of strict domestic control and enhanced international vigilance to combat diversions into the illicit traffic.[9]

Attempts to enforce the provisions of the 1925 International Opium Convention demonstrated the inadequacies of the control regime. Although national statistical returns provided a clearer picture of world supply and demand and many governments strengthened domestic enforcement measures, new problems arose. As Western European governments pressured pharmaceutical companies to conform to more stringent control standards, unscrupulous operators moved to jurisdictions that had not ratified the 1925 treaty. Traffickers became more sophisticated in their operations, colluding with political and/or military power brokers to avoid prosecution. Most observers agreed that chaotic conditions in China and inadequate enforcement in agricultural producing countries such as Iran and Turkey made it impossible to enforce significant limitation of excess agricultural production. Moreover, use of manufactured substances appeared to increase, especially in East Asia. Western control officials feared that abuse of substances such as morphine, codeine, and cocaine might rise drastically in industrialized states.

Consequently, by 1931 many governments concluded that further treaty negotiations were necessary to fight the drug problem. Attention focused on schemes to curb surplus manufacture. Yet a deepening worldwide depression profoundly affected the negotiations. Agricultural producing states did not want to hinder exports of important cash crops such as opium and coca. Governments housing significant pharmaceutical firms also wished to protect their ability to earn foreign currency through manufacture and export of medicines. Consuming countries wished to acquire vital medicaments at affordable prices. As a result, delegates, including Commissioner Anslinger, from industrialized Western states that housed significant pharmaceutical manufacturing industries rejected proposed quota schemes that would have limited competition by assigning each company a certain percentage of the licit trade.

The 1931 conference instead settled on indirect measures to control the trade in manufactured drugs. The treaty called for signatories to estimate, one year in advance, their requirements for opiates and coca products. Governments also required pharmaceutical firms to estimate how much of each controlled substance they expected to manufacture. The treaty created a new control organ, called the Drug Supervisory Body, to examine those estimates and publicize any discrepancies or cases that might indicate leakage into the illicit traffic. The provisions of the 1925 treaty for import/export authorizations were extended to manufactured substances, enabling the Permanent Central Opium Board to track the worldwide legal trade of all significant addicting drugs. After considerable discussion, influenced by representatives of pharmaceutical manufacturing firms, the convention decided to create a less stringent control level (known as Schedule II) for codeine and similar drugs widely used in medical practice and generally considered a lower abuse risk.[10]

Taken together, the 1925 and 1931 treaties created a drug control regime based on several key tenets:[11]

Supply control became enshrined as the ruling paradigm of the regime. The goal of the system was not to eliminate availability altogether, but rather to reduce supply to medicinal, scientific, and industrial needs. Many observers have misunderstood the thrust of international/national legislation by confusing regulation with prohibition. Prohibition has rarely been enacted because almost all substances retain some legitimate uses. Rather, control advocates religiously adhered to the contention, however questionable, that by drying up excess capacity that might otherwise migrate to the illicit market, the addiction problem would disappear of its own accord.

National control retained pride of place at the expense of substantive restrictions imposed by external agencies. Nation-states proved unwilling to surrender their prerogatives. Consequently, governments circumscribed the powers of supranational regulatory bodies, such as the Permanent Central Opium Board, created by international treaty.

Indirect (as opposed to direct) control was adopted. Governments agreed to report estimates of need, actual usage, imports, exports, and reserve stocks to the international agencies. The international authorities received no power to approve transactions ahead of time (direct control), but only to object after the fact to any behavior that appeared inappropriate (indirect control).

The international regime favored *free trade* over substantive limitations on manufacture and/or agricultural production. Attempts to institute quotas for production, manufacture, and/or consumption consistently failed. This consideration played an especially important role during the worldwide depression of the 1930s. Consumers did not want to pay more than necessary for legitimate medicinal products. Neither agricultural states that produced raw materials nor manufacturing states that made finished pharmaceutical products wanted significant limitations on their freedom to sell in the legitimate market. As a consequence, the system featured considerable excess supply and capacity, and thus attempted to regulate the licit trade while suppressing illegitimate traffic.

Control was accomplished through *schedules* based on presumptions about addictive propensity. Control officials concluded that some substances, such as codeine, should not be limited by the same strictures as more addicting and less medicinally useful substances such as heroin. Narcotics, coca products, and marijuana were generally considered guilty until proven innocent, although the opposite calculation applied to psychotropics developed after World War II. Regulatory mechanisms provided medical experts, pharmaceutical companies, the research community, and industrial interests with opportunities to state their positions during the scheduling process, but political appointees and bureaucrats rather than technical experts usually make the key scheduling decisions. Since its introduction in the 1931 treaty, this "schedule approach" to control, featuring tiered levels of regulation, influenced the research and development agendas of pharmaceutical manufacturers. Companies strove to concoct medicinally useful substances that did not meet the extant criteria for addiction. In so doing, fundamental questions about the nature of addiction and why individuals chose to abuse remained unstudied and unanswered.

In sum, the configuration of the regime in place by the mid-1930s amounted to a struggle for *comparative regulatory advantage* within a set of relatively predictable worldwide marketing rules. The framers intended to create a barrier high enough to keep out unscrupulous players but low enough not to impede commerce among "legitimate" firms. The principal emphases of importance were that the international regulatory scheme applied to drugs existed in an atmosphere that emphasized supply control at the expense of other policy options such as rehabilitation, treatment, and/or prevention; that the regime was designed primarily to control rather than to prohibit; and that geopolitical and other concerns were more important than medical factors in the creation and operation of the system.

The 1931 treaty also provided a great benefit to Anslinger's domestic position: he utilized one of its clauses to claim that international treaty obligation required the U.S. government to maintain his agency. Anslinger used this questionable assertion repeatedly to provide a bulwark against bureaucratic reorganization proposals that would have eliminated the FBN and possibly resulted in different drug control policies in the United States.

MID-1930S TO MID-1960S: IMPLEMENTING THE RULES

Until his retirement in 1962, Anslinger attempted to apply these strictures in a combination that would minimize domestic abuse, maximize American pharmaceutical firms' capacity to sell overseas, and curb excess agricultural production in far-flung areas of the globe.

Yet it quickly became evident to Anslinger that his international goals could not soon be achieved. The Sino-Japanese war beginning in 1931 and chaotic conditions within China contributed to an uncontrollable East Asian drug traffic. A growing threat from Nazi Germany portended potential war in Europe.

Moreover, from Anslinger's point of view, international negotiations failed him again in the mid-1930s. After the 1925 and 1931 treaties created a basic control system that instituted restrictions on manufacture and trade, it became increasingly evident that the new rules had fostered the growth of a large clandestine drug market. Consequently, the League of Nations attempted to deal with the issue of transnational drug trafficking. In 1936, forty-two governments convened to negotiate an illicit drug trafficking convention. They could not forge an international consensus, however, because states differed on what drugs should be included in the treaty, which acts should be defined as offenses, and how enforcement measures, such as extradition, should be implemented. Reconciling differences in national legal systems also presented a serious stumbling block. The Americans attempted to promote an aggressive agenda, including criminalizing all agricultural production not necessary for medical purposes and quasi-medicinal forms of drug use such as smoking opium. Many colonial powers could not accept such provisions because they would have defined many Asian subjects as offenders. Frustrated, the American delegation withheld active participation and privately disparaged the proceedings. Although delegates did produce a treaty, its provisions proved too general for most pro-control governments and too specific for those wishing to avoid further obligations. The 1936 treaty never gained widespread acceptance. Those governments interested in pursuing traffickers now negotiated separate bilateral agreements with like-minded states.

Anslinger's domestic pronouncements and policies during the 1930s must be seen within this larger context. The commissioner feared, on the one hand, that an explosion of excess supply might swamp his domestic law enforcement efforts. Any surrender to less stringent enforcement, weaker laws, ambulatory treatment, or medicalization, he believed, would ultimately harm American citizens. He also wanted to avoid the situation the United States found itself in during World War I, when the country possessed inadequate medical supplies to fight a protracted war. Consequently, he acted to ensure reasonable prices for licit supplies in order to ensure an adequate stockpile for whatever contingency might occur.[12]

Based on this wider perspective, a key to understanding the most important legislation of the pre–World War II decade, the 1937 Marihuana Tax Act, lies in Anslinger's view of the post-1933 geopolitical situation.[13] Japan attacked Manchuria in 1931, precipitating an international crisis and an ever-deepening war in East Asia. Hitler came to power in January 1933 and soon his bellicose statements and actions aroused concerns in Western capitals. By the mid-1930s, real threats to peace appeared across both America's Atlantic and Pacific shores.

Consequently, the War Department created and continually updated a list of critical and strategic raw materials, essential to the defense of the nation, that could not be secured easily from domestic sources. Hemp was included on the list because it served as the only viable substitute for several tropical hard fibers (jute, sisal, silk, manila fiber, henequen), the supply of which might be cut off in wartime.[14]

Once it became clear that the government might have to institute a crash program to grow hemp in the United States, Commissioner Anslinger altered his position on marijuana legislation. He had previously advocated prohibition, but in 1935-1936 changed to a control-oriented approach. Anslinger had to make hemp available for industrial purposes, yet he feared that illicit use of marijuana might create clandestine pathways that smugglers could use to traffic in other drugs. Consequently, he constructed legislation that separated the resin that contemporary experts believed contained the psychoactive ingredient from the inert industrial product. The 1937 Marihuana Tax Act empowered the federal government to supervise marijuana production, distribution, and industrial use by licensing those who handled the substance. To purchase the proper tax stamp, farmers and manufacturers had to register with the FBN and account for seeds received, crop harvested, and disposal of leaves, tops, and stems.[15]

The principal reason that this international motivation behind the Marihuana Tax Act was not discussed openly at the time also related to the global geopolitical situation. The government feared an open admission of its war-related stockpiling efforts would upset domestic pacifists and isolationist supporters. Washington also wished to deter speculators from bidding up the price of strategic commodities. Finally, responsible officials did not want to provide information to potential enemies about America's weaknesses. Taking these extranational considerations into account helps to explain the timing and configuration of the inaugural American cannabis control statute by emphasizing the mind-set that prevailed at the time of the tax act.[16]

Most contemporaries concluded that Anslinger's precautions proved warranted; during World War II the federal government instituted an emergency War Hemp Program that put hundreds of thousands of acres, primarily in Kentucky, Illinois, Wisconsin, and Minnesota, into hemp production. The commissioner's foresight ensured an enhanced fiber supply while preventing any apparent increase in illicit drug trafficking and use.[17]

During World War II and the immediate postwar years, Anslinger engaged in several major international initiatives. The FBN not only had important defense-related tasks to fulfill, but the commissioner also viewed the war as an opportunity to achieve several related long-term goals. He hoped to accomplish what American diplomacy had failed to do in peacetime—create a comprehensive, global drug control regime in accordance with the American approach: strict control over agricultural production, careful monitoring of trade, manufacture, and distribution, and stringent domestic enforcement against illicit use. The supply control paradigm remained intact; American authorities intended to curb drug abuse by eliminating the leakage of licit supplies into the illicit traffic. Anslinger's ability to use American power as leverage to exact compliance from other states was at its height between the early 1940s and the early 1950s. Ultimately he sought to create an international environment that complemented his efforts at home.

Anslinger's first defense-related objective focused on maintaining an adequate supply of drug-related raw materials and manufactured drugs necessary to meet the medicinal needs of the United States and its allies. He oversaw a massive stockpiling program that he hoped would suffice. Nevertheless, as insurance against disruption of supply lines, Anslinger authorized confidential plans that would enable the government to grow opium in the United States and coca in Puerto Rico. As part of that program, he quietly promoted legislation that would enable the FBN to regulate agricultural production of addicting substances on U.S. soil. As a result, the FBN occupied a commanding position: it controlled legal drug supplies for most of the free and neutral world. When foreign governments pleaded for desperately needed medicaments, Anslinger insisted on compliance with international control regulations before he would authorize U.S. pharmaceutical firms to fill orders.

To promote the war effort, the commissioner also tried to deny Germany access to important medicines, but in the process he became disillusioned with several key allies. He gathered all available information about German imports, production capabilities, manufacturing capacity, and the development of synthetic alternatives to opium-based medicines. He promoted aggressive measures to reduce excess agricultural production in Allied-controlled territory and advocated stringent restrictions on shipments of addicting medicines, arguing that only substantial need merited the risk of sending drugs overseas because they might be diverted into Axis hands. Yet American allies proved, in Anslinger's eyes, less than cooperative. British pharmaceutical firms attempted to sell more narcotics abroad, especially in South America, much to the consternation of control authorities and American pharmaceutical manufacturers who had entered into that traditionally British trade preserve. Furthermore, Great Britain refused to eliminate the substantial opium cultivation in India that supported quasi-medicinal use in the colony and also supplied the raw material for the English pharmaceutical companies' trade offensive. Upon reoccupying North Africa, the French reinstituted hashish monopolies that sold addicting substances and profited the local administration. The French and Dutch governments appeared poised to reinstate opium distribution monopolies, originally instituted around the turn of the century as a reform measure, in their Asian colonies. Those government-sponsored arrangements for selling opium to licensed users had long drawn the ire of American control advocates. The commissioner concluded that allied governments were not serious about enacting stringent controls and that he would have to take more coercive measures to achieve his goals.

With assistance from his Canadian counterpart, Colonel C. H. L. Sharman, Harry Anslinger used a combination of bluff and bluster to pressure the British, Dutch, and French governments into repudiating their East Asian colonial opium monopolies. By eliminating the government-sponsored selling of opium to legally registered users, Anslinger believed he had eliminated one of the principal sources of excess supply and one of the major conduits for illicit trafficking.

In conjunction with several key American supporters and international officials, during 1944-1945 Harry Anslinger promoted a vigorous drug control apparatus as governments contemplated how to organize the United Nations. The UN created the Commission on Narcotic Drugs, comprising governmental representatives, to oversee international drug control efforts. Anslinger successfully lobbied many governments to appoint representatives who shared his strong supply control orientation. The commissioner also installed like-minded officials in several key positions on the Permanent Central Opium Board, the Drug Supervisory Body, and in the UN secretariat.[18]

With allies occupying many important positions, Anslinger in the later 1940s spearheaded a worldwide campaign to implement his supply control vision. Officials resurrected regulatory mechanisms in war-torn territories, paying special attention to the former Axis nations. Anslinger also participated in a successful effort to forge a 1948 protocol that placed synthetic narcotics under the same manufacturing and trade restrictions as those drugs derived from the poppy plant. The opium monopolies of the Asian colonial administrations expired gradually. Control officials also suppressed several initiatives to introduce licit opium production and manufacturing in Latin American states.

Yet those efforts only maintained the extant level of control. The attempts of Anslinger and like-minded officials to impose substantive supply control on a global basis proved disappointing. The Nationalist Chinese, driven from power by the Communists, stimulated opium production in China's southern border areas to pay for continued operations. Iran emerged as a massive supplier to the illicit market, and excess production in Turkey could only be constrained through considerable diplomatic pressure. Burma, Laos, and Thailand became centers of a new and frightening illicit manufacturing nexus, concocting street-salable heroin in clandestine laboratories. Drug trafficking on a worldwide scale grew steadily. Political and economic considerations hindered control advocates' efforts; leading Western governments often ignored drug trafficking in third world states because they did not want to endanger relations with allies in the Cold War fight against Communism. By 1948-1949 Anslinger concluded that his piecemeal efforts to impose substantive supply control had largely failed.

The commissioner then moved to create a new international treaty intended to supersede all previous agreements, but this effort also failed. Dubbed the "Single Convention," Anslinger hoped to create a document that would impose a uniform stringent drug control regime worldwide. His efforts, however, were derailed by a competing effort emanating from the UN bureaucracy. After several years of internecine fighting, the 1953 Opium Protocol emerged instead of the Single Convention. A 1953 protocol created a cartel arrangement in which a few agricultural producing countries monopolized legal opium sales in exchange for their strict policing of the illicit trade. Manufacturing firms and governments had to purchase their raw materials from this closed list of suppliers. The treaty also provided for independent inspections of a nation's territory to confirm that no clandestine cultivation occurred. Many agricultural producing countries and communist governments objected to that provision because it infringed on national sovereignty. The agreement did not include provisions for eliminating illicit cultivation and trafficking from other quarters. Nor did it deal with other potentially problematic drugs such as synthetic narcotics, coca products, or cannabis. Because of its omissions and controversial provisions, the 1953 protocol divided rather than united the world's governments. Many officials began to question the tenets of the supply control approach. Even if the Opium Protocol could be enforced, it was not clear that a strict supply-side effort could "solve" the "drug problem." Not only did Anslinger's prestige and domestic bureaucratic standing suffer as a result, he was once again frustrated in his attempts to secure decisive cooperation in the international arena.[19]

This international perspective helps explain many of Anslinger's postwar domestic maneuvers. He blamed the Communist Chinese for much of the world's drug problem, despite a lack of evidence, because accusing Beijing enabled the FBN to insinuate itself into the burgeoning Cold War-intelligence-defense complex. Accusations about foreign encroachment also benefited the FBN in intergovernmental jurisdictional battles. The Customs Service was assigned drug control duties in the Far East while the FBN handled Europe; Anslinger intended to demonstrate the superiority of his agents in hopes of gaining more personnel and authority. The publicity about Mafia influence over the drug traffic challenged FBI Director J. Edgar Hoover's repeated pronouncements that no significant Mafia activity existed in the United States, strengthening Anslinger's hand vis-à-vis his better-known rival.

Anslinger's support for punitive American legislation also reflected his international perspective and concerns as discussed in Chapters 3 and 4. The 1951 Boggs Act imposed much stiffer penalties for first convictions and ensuing violations, and eliminated the courts' options to apply suspended sentences or probation on subsequent convictions. The 1956 Narcotic Control Act drastically increased minimum mandatory sentences again, imposed higher fines, and provided for the death penalty in cases of adults convicted of selling morphine to minors. The commissioner supported heavy penalties in part because he believed it important to provide an example for other governments—if the United States faltered in its commitment to demand severe penalties, Anslinger could hardly urge other nations to tighten their laws. Similarly, in Anslinger's eyes any surrender to ambulatory treatment, medicalization of addicts, or education efforts might give other nations an excuse to avoid implementing serious control measures. Most important, as it became increasingly apparent in the early 1950s that the United States could not secure sufficient international cooperation to eliminate the "drug scourge," Anslinger believed his only resort was to tighten domestic controls. By the late 1950s Anslinger and the FBN, having run out of ideas, were reduced to a defensive reiteration of policies recited since the 1930s.

Similarly, the opposition to the draconian policies championed by Anslinger that grew slowly during the 1950s reflected a growing awareness of the international scene. Those who questioned FBN policy, including organizations such as the American Bar Association, the American Medical Association, and knowledgeable individuals such as highly regarded sociologist Alfred Lindesmith, did so in part because they concluded that international negotiations would never provide a resolution to the domestic drug problem. Other governments and the UN searched for policy alternatives to a continued emphasis on policing, including efforts to enhance regional cooperation and the first serious consideration of crop substitution and/or foreign development aid.[20]

LATE 1960S TO 2000: BRAVE NEW WORLD

The decade of the 1960s marks a great watershed in drug history. The illicit use of drugs traditionally targeted by the control regime (heroin and other opiates, cocaine, and cannabis) expanded and spread into "respectable" middle-class communities. Concerns over expanding consumption of alcohol also increased during this decade. Moreover, use of psychotropic substances (stimulants, depressants, and hallucinogens) multiplied rapidly, leading to heightened concerns about the abuse potential of these newer substances. Notably, this explosion of demand for drugs occurred on a global scale; a significant rise in substance abuse occurred in industrialized states, developing countries, and apparently behind the Iron Curtain as well.[21]

During the period coinciding with the Kennedy and Johnson administrations, the control apparatus responded largely according to the established pattern, concentrating on controlling supplies of the "traditional" drugs of abuse. Because the 1953 Opium Protocol did not receive widespread support, attention returned to the project of forging a unified treaty that would consolidate the complex webs of obligations and close the loopholes created by the nine extant international agreements.[22]

That effort culminated in the UN-sponsored 1961 Single Convention, which incorporated the key provisions of previous treaties into one document. It required governments to submit estimates of need in advance and collect statistics concerning cultivation, manufacture, trade, and stockpiling of opiates, coca products, and cannabis. Existing import/export certification procedures remained intact. The functions of the Permanent Central Opium Board and the Drug Supervisory Body were combined into the International Narcotics Control Board, which examined reports and publicized discrepancies. The schedules of control that afforded differing levels of regulation were retained and refined by the creation of four separate schedules. The Single Convention also eliminated the most objectionable aspects of the 1953 Opium Protocol, including the provisions for independent inspections and the closed list of legitimate suppliers. Those omissions enabled governments to retain their national sovereignty and allowed the free market to remain intact. Most important for Harry Anslinger, the 1961 pact also posed a threat to the FBN because it did not incorporate language from previous agreements that Anslinger used to support the independence of his agency.[23]

Between 1961 and 1967, an international dispute concerning whether the 1953 Opium Protocol or the 1961 Single Convention should serve as the basis for international regulatory arrangements preoccupied control officials. Typical of drug diplomacy, the conflict involved calculations about political, economic, strategic, and other nonmedical considerations. Although Anslinger retired from the FBN in 1962, he continued to serve as U.S. representative to the Commission on Narcotic Drugs (CND), the key UN body responsible for making international control decisions. Anslinger lobbied for the 1953 Opium Protocol, the more restrictive of the two treaties because it focused on severely limiting opium cultivation and supply. He also opposed the 1961 Single Convention because it only amalgamated the major provisions of several existing treaties without breaking new regulatory ground. Finally, he wished to protect the FBN from bureaucratic rivals. In 1962 the commissioner won an intragovernmental battle with the State Department over which agreement the United States would officially support. As had happened many times previously, Anslinger and his bureaucratic successors advocated no reduction of strict domestic control measures, both to reinforce American influence with wavering overseas governments and also to serve as a firebreak in case other nations rejected his enforcement-oriented approach. Anslinger launched an aggressive international campaign on behalf of the unpopular 1953 protocol, earning the ire of many in diplomatic circles at home and abroad. He employed the same sort of high-pressure tactics he had used many times before: Anslinger's and the FBN's reputation for "bomb-throwing" on the international diplomatic scene dated back to the 1930s and had often caused embarrassment for other branches of the U.S. government. After 1964 the State Department gradually recovered its control over foreign drug policy, signaled by U.S. ratification of the Single Convention in 1967. Until the later 1960s, the internecine battles that hamstrung U.S. international policy reflected the ambivalence of American policymakers about what course to follow with respect to the drug question.

As the "drug problem" reached unprecedented worldwide proportions in the late 1960s and into the 1970s, a new generation of policymakers in the United States and abroad encountered the continuing international factors that impeded progress toward a solution. Ideas such as those proposed by longtime FBN agent John Cusack that focused on pressuring key agricultural producing states had certainly been tried before. Two complications, however, diminished the efficacy of such exercises. Countervailing political and economic pressures often weakened pressure on target states. For example, U.S. officials often looked the other way concerning drug trafficking because they did not want to endanger foreign governmental cooperation in prosecuting the Cold War. Moreover, even when pressure tactics did achieve a significant reduction in illicit activity, this business simply moved elsewhere. Thus, after the United States compelled Turkey to comply with internationally imposed expectations in the early 1970s, traffickers and agricultural producers merely moved to

Mexico, then South America and Asia. Control advocates repeatedly failed because they could neither overcome competing policy imperatives nor exert sufficient enforcement in all parts of the globe simultaneously. The Nixon and subsequent administrations learned what previous policymakers had concluded: in order to present a united front against trafficking and illicit use, it was necessary to engage in the messy business of reaching global agreement through international treaty negotiations. Many other governments, beset by similar burgeoning drug problems, concluded much the same.

Yet those responsible for enforcing drug control reacted especially slowly regarding the increasing problem with the new psychotropic drugs. Most Western governments did not impose restrictions on the availability of psychotropic substances until the later 1960s. International action took even longer. Some governments had attempted since the early 1950s to expose the dangers posed by these new substances. Scandinavian countries, for example, experienced early difficulties with amphetamine and barbiturate abuse, and objected to the unrestricted export policies of manufacturing states such as Germany and the Netherlands. Concerned governments directed their comments to the Commission on Narcotic Drugs, the UN body comprising of governmental representatives charged with overseeing international drug control efforts. Yet the CND, led by Western governments with influential pharmaceutical industries, discouraged efforts to place the issue on the international agenda.[24]

A treaty designed to regulate psychotropic substances was finally negotiated in 1971, but the regime it instituted called for much less stringent controls than those placed on opiates and cocaine. The 1971 Psychotropic Convention, for example, did not include any provision for estimates of need, a key element in assessing whether pharmaceutical firms were manufacturing excess (i.e., nonmedicinal) supplies. Nor did the treaty include the usual language incorporating derivative substances (the salts, esters, ethers, and isomers and other preparations that comprised the bulk of medicinal applications) in the scope of control. The 1971 pact featured less thoroughgoing reporting requirements, included relatively few substances in each of its four schedules, and had no requirement to track precursor chemicals. In general, the 1971 convention treated psychotropic drugs as "innocent until proven guilty"—exactly the opposite of the formula long employed to control opiates, synthetic narcotics, coca products, and cannabis.[25]

Psychotropics received more lenient treatment for several reasons. Pharmaceutical companies and the governments that represented them acted vigorously to protect commercial opportunities at the expense of more stringent control measures. Other practical difficulties complicated matters. Those who negotiated national legislation and international treaties recognized that the range of substances in question was much larger, with a much wider range of therapeutic applications. No one wished to discourage responsible use of psychotropics, and as a consequence most contemporaries believed the control regime's design should demonstrate somewhat more tolerance than that devised for opiates and coca products.

Nevertheless, psychotropics received, and to a certain extent still receive, a largely undeserved "pass" at both the national and international levels. Subsequent experience has continually demonstrated that a fairly sizable number of "wonder drugs" have not produced the sort of addiction-free benefits claimed by their supporters.[26]

Two additional factors contributed significantly to this slow, inadequate response to the danger posed by psychotropics. The problem, in part, was one of definition. The existing drug control treaties into the 1960s defined an addicting substance as one that generated effects similar to those produced by opiates or coca products. Central nervous system stimulants, depressants, and hallucinogens acted differently upon the body; ergo, many assumed they must not be addictive. The physical and psychological manifestations of psychotropic abuse forced the medical and treatment communities to reconsider some of their key concepts, a process that took time to evolve. Moreover, the officials running the control system, be they government bureaucrats, medical practitioners, scientific researchers, or pharmacological experts, all operated within a culture that adopted a more indulgent attitude toward psychotropics because they were the products of Western science. People in the West generally felt more familiar with and accepting of psychotropics precisely because they emerged from a process of scientific experiment carried out by highly qualified experts imbued with authority, using a scientific process refined over a century. This cultural predisposition to view psychotropics favorably parallels the permissive treatment afforded alcohol—there has never been any serious attempt to impose international control over that substance. Opiates and cocaine remain, to Westerners, comparatively alien drugs, associated with undesirable user groups. Efforts to control those substances produce less resistance, largely because they do not have socially respectable and economically powerful defenders.[27]

The drug demand explosion beginning in the 1960s permanently altered the trajectory of international drug control efforts. For the first time officials paid serious, sustained attention to issues of demand. In addition to governments, national and international medical associations, health agencies, social service organizations, judicial authorities, and other interested parties explored a variety of prevention, intervention, treatment, and enforcement alternatives to the dominant control and enforcement paradigm. Governments and international agencies supported development aid and crop substitution programs in hopes of undermining the economic incentives to grow illicit substances. Governments also pursued regional cooperative approaches to law enforcement. Within the United States, alternatives included a variety of treatment modalities, including methadone maintenance, more sophisticated intervention procedures, initiatives designed to prevent or delay first drug use among youth, and calls for decriminalization. This programmatic dispersion produced many promising new avenues to deal with the drug question, but it also diffused the previous societal consensus on the nature and scope of the problem and how best to confront it. Since the later 1960s, Western industrialized countries, in particular, debated vigorously whether the primary responsibility for drug abuse should be traced to individual pathology, familial dysfunction, economic deprivation, legislative inefficacy, or societal malady.

At the same time, largely owing to the lack of consensus on alternative ameliorative conceptions, the supply control paradigm remained robust. Governments and nongovernmental agencies continued to focus the majority of their efforts and resources on restricting supplies, interdicting illicit trafficking, and maneuvering for advantage with respect to the legal drug trade. The governments of countries housing the leading pharmaceutical manufacturing companies continued resisting placing increased restrictions on psychotropics.

A 1988 UN-sponsored antitrafficking treaty serves as an indicator of late-twentieth century changes and continuities. Unlike the 1936 effort, this treaty received widespread cooperation and acceptance, a sign of the seriousness with which governments worldwide now viewed the drug problem. The 1988 Illicit Traffic Convention criminalized illegal trading of precursor chemicals (a clause governments rejected in 1971), money laundering, and international trafficking. The treaty also required parties to pass legislation allowing for the confiscation of offenders' assets and it called upon governments to cooperate in law enforcement matters. Signatories agreed to expedite extradition in certain circumstances, always a touchy subject with governments jealous to protect national sovereignty. The 1988 treaty even included a provision, albeit weak, calling for the eradication of illicit cultivation. Some important governments proved reluctant to ratify this treaty at first, but resistance ebbed over time as the treaty's strictures became more widely accepted.[28]

By the early 1990s UN pronouncements and activities exemplified this dual-track approach to the drug problem. A 1988 UN-sponsored conference created the Comprehensive Multidisciplinary Outline of Future Activities in Drug Abuse Control (CMO). The CMO placed equal stress on demand reduction and prevention, supply control, suppression of illicit trafficking, and treatment/rehabilitation. In addition to giving issues other than supply serious attention, the CMO recognized many of the problems associated with abuse of psychotropic substances. In 1990 the UN devised a System-Wide Action Plan that attempted to coordinate the activities undertaken by UN agencies and associated organizations. The UN General Assembly held a special session in 1990 devoted to the drug issue and declared 1991-2000 as the United Nations Decade Against Drug Abuse.[29] What influences these programs and pronouncements will have over the long term remains to be seen, but it appears unlikely that any solution to the drug problem is imminent.

CONCLUSION

Since the late nineteenth century, the American drug experience has largely mirrored that of other Western industrialized nations. The United States has acted as a center of demand for licit drugs and illegal substances, as well as for regulatory activism. The United States has served as a principal engine propelling developments in drug technology, jurisprudence, and enforcement. Domestic initiatives have inevitably been affected by, and have reflected upon, international drug control activities. Despite an oft-expressed desire to eschew the international complications of the drug question, policymakers, legislators, and citizens of the United States, much like addicts, cannot escape their relationship to the global drug scene.

APPENDIX: MULTILATERAL TREATIES ON NARCOTICS AND PSYCHOTROPIC SUBSTANCES

For the text of several of the treaties listed here and other information related to international drug control, consult the following Web sites:

- **1912 Hague Opium Convention:** International Opium Convention, signed at The Hague on January 23, 1912*

  <http://untreaty.un.org>
  <http://www.incb.org>
  <http://www.unodc.org/unodc/index.html>

- Agreement Concerning the Manufacture of, Internal Trade in and Use of Prepared Opium, signed at Geneva, February 11, 1925

- **1925 Geneva Opium Convention:** International Opium Convention, signed at Geneva on February 19, 1925*

- **1931 Geneva Manufacturing Convention:** Convention for Limiting the Manufacture and Regulating the Distribution of Narcotic Drugs, signed at Geneva on July 13, 1931*

- **1931 Bangkok Agreement:** Agreement for the Control of Opium Smoking in the Far East, signed at Bangkok on November 27, 1931*

Note: Treaties mentioned in this chapter are indicated with an asterisk (*); unofficial titles by which the treaties are most commonly known are in bold type.

- **1936 Illicit Trafficking Convention:** Convention for the Suppression of the Illicit Traffic in Dangerous Drugs, signed at Geneva on June 26, 1936*

- **1946 Protocol:** Protocol signed at Lake Success on December 11, 1946, amending the Agreements, Conventions and Protocols on Narcotic Drugs concluded at The Hague on January 23, 1912; at Geneva on February 11, 1925, February 19, 1925, and July 13, 1931; at Bangkok on November 27, 1931; and at Geneva on June 26, 1936

- **1948 Protocol:** Protocol Signed at Paris on November 19, 1948, Bringing Under International Control Drugs Outside the Scope of the Convention of July 13, 1931, for Limiting the Manufacture and Regulating the Distribution of Narcotic Drugs, as Amended by the Protocol Signed at Lake Success on December 11, 1946

- **1953 Opium Protocol:** Protocol for Limiting and Regulating the Cultivation of the Poppy Plant, the Production of, International and Wholesale Trade in, and Use of Opium, signed at New York on June 23, 1953*

- **1961 Single Convention:** Single Convention on Narcotic Drugs, signed at New York on March 30, 1961*

- **1971 Psychotropic Convention:** Convention on Psychotropic Substances, signed at Vienna on February 21, 1971*

- **1972 Protocol Amending the 1961 Single Convention:** Protocol Amending the Single Convention on Narcotic Drugs, 1961, signed March 25, 1972

- **1988 Illicit Trafficking Convention:** Convention Against Illicit Traffic in Narcotic Drugs and Psychotropic Substances, signed at Vienna on December 20, 1988*

- General UN Drug Control Program (UNDCP)

NOTES

1. Joseph Spillane, *Cocaine: From Medical Marvel to Modern Menace in the U.S., 1884-1920* (Baltimore: Johns Hopkins University Press, 2000).

2. Steven B. Karch, *A Brief History of Cocaine* (Boca Raton, FL: CRC Press, 1998); Paul Gootenberg (Ed.), *Cocaine: Global Histories* (London: Routledge, 1999); Marcel de Kort and Dirk J. Korf, "The Development of Drug Trade and Drug Control in the Netherlands: A Historical Perspective," *Crime, Law, and Social Change* 17 (1992): 123-144.

3. Jonathan Spence, "Opium," *Chinese Roundabout: Essays in History and Culture* (New York: Norton, 1992), pp. 228-258.

4. P. J. Giffen, Shirley Endicott, and Sylvia Lambert, *Panic and Indifference: The Politics of Canada's Drug Laws* (Ottawa: Canadian Centre on Substance Abuse, 1991); Norman E. Zinberg, *Drug, Set, and Setting: The Basis for Controlled Intoxicant Use* (New Haven: Yale University Press, 1984); Allan M. Brandt and Paul Rozin (Eds.), *Morality and Health* (New York: Routledge, 1997), especially David T. Courtwright, "Morality, Religion, and Drug Use," pp. 231-250; Roy Porter and Miklaus Teich (Eds.), *Drugs and Narcotics in History* (Cambridge: Cambridge University Press, 1995), Chapters 4, 5, 6; Marcel de Kort, "Drug Policy: Medical or Crime Control? Medicalization and Criminalization of Drug Use, and Shifting Drug Policies," in Hans Binneveld and Rudolph Dekker (Eds.), *Curing and Insuring: Essays on Illness in Past Times: The Netherlands, Belgium, England, and Italy, 16th-20th Centuries* (Hilversum: Verloren, 1993); de Kort and Korf, "Development of Drug Trade"; Virginia Berridge, "Drugs and Social Policy: The Establishment of Drug Control in Britain, 1900-30," *British Journal of Addiction* 79 (1984): 17-29; Terry M. Parssinen, *Secret Passions, Secret Remedies: Narcotic Drugs in British Society, 1820-1930* (Philadelphia: Institute for the Study of Human Issues, 1983); H. Richard Friman, *Narcodiplomacy: Exporting the U.S. War on Drugs* (Ithaca, NY: Cornell University Press, 1996).

5. Arnold H. Taylor, *American Diplomacy and the Narcotics Traffic* (Durham: Duke University Press, 1969); David Musto, *The American Disease* (Oxford: Oxford University Press, 1987); Peter D. Lowes, *The Genesis of International Narcotics Control* (Geneva: Librarie Droz, 1966); David Edward Owen, *British Opium Policy in China and India* (New Haven: Yale University Press, 1934); Charles Clarkson Stelle, *Americans and the China Opium Trade* (New York: Arno Press, 1981); F. S. L. Lyons, *Internationalism in Europe, 1815-1914* (Leyden: A. W. Sijthoff, 1963); Warren F. Kuehl, *Seeking World Order: The U.S. and International Organization to 1920* (Nashville: Vanderbilt University Press, 1969).

6. Virginia Berridge, "War Conditions and Narcotics Control: The Passing of Defence of the Realm Act Regulation 40B," *Journal of Social Policy* 7(3) (1978): 285-304; Giffen, Endicott, and Lambert, *Panic and Indifference;* Judy Slinn, "Research and Development in the U.K. Pharmaceutical Industry from the Nineteenth Century to the 1960s," in Porter and Teich, *Drugs and Narcotics in History,* pp. 168-176; Caroline Acker, "Addiction and the Laboratory: The Work of the National Research Council's Committee on Drug Addiction, 1928-1939," *Isis* 86 (1995): 167-193; John P. Swann, *Academic Scientists and the Pharmaceutical Industry: Cooperative Research in Twentieth-Century America* (Baltimore: Johns Hopkins Uni-

versity Press, 1988); Victoria A. Harden, *Inventing the NIH: Federal Biomedical Research Policy, 1887-1937* (Baltimore: Johns Hopkins University Press, 1986).

7. Kathryn Meyer and Terry M. Parssinen, *Webs of Smoke: Smugglers, Warlords, Spies, and the History of the International Drug Trade* (Lanham, MD: Rowman and Littlefield, 1998).

8. Taylor, *American Diplomacy;* Warren F. Kuehl and Lynne K. Dunn, *Keeping the Covenant: American Internationalists and the League of Nations, 1920-1939* (Kent, OH: Kent State University Press, 1997); F. P. Walters, *A History of the League of Nations* (London: Oxford, 1952); *The League of Nations in Retrospect: Proceedings of the Symposium* (Berlin: Walter de Gruyter, 1983). For documents relating to the 1924-1925 conferences, see: League of Nations documents C.760. M.260.1924.XI. and C.684.M.244.1924.XI.; Raymond Leslie Buell, "The International Opium Conferences with Relevant Documents," *World Peace Foundation Pamphlets* 7(2-3) (1925): 39-330; British delegation report, May 13, 1925, Foreign Office, *The Opium Trade, 1910-1941,* F.O. 415: Correspondence Respecting Opium, Volume XXII (January-December 1925), No. 13, Public Record Office (Wilmington, UK: Scholarly Resources, 1974).

9. Record of conference, October 7, 1930, Drug Enforcement Administration archives, ACC 170-71-A, Box 10, File 0355 (Geneva Convention); Oscar Ewing, "The Narcotic Battle at Geneva," 1931 (undated), Anslinger Papers, Pennsylvania State University, Box 10, File 11.

10. Taylor, *American Diplomacy.* For records relating to the 1931 conference see: League of Nations document C.509.M.214.1931.XI.

11. League of Nations Historical and Technical Study, document C.191.M.136. 1937.XI.

12. William B. McAllister, *Drug Diplomacy in the Twentieth Century: An International History* (London: Routledge, 2000) Chapter 4; Gary B. Ostrower, *Collective Insecurity: The U.S. and the League of Nations in the Early Thirties* (Lewisburg, PA: Bucknell University Press, 1979).

13. William B. McAllister, "The International Connection: How Visions of War and the Dilemmas of Strategic Planning Shaped the Marihuana Tax Act of 1937," lecture, University of Michigan, Substance Abuse Research Center, March 1999.

14. Army and Navy Munitions Board, *The Strategic and Critical Materials,* hearings on HR 2969, 3320, 2556, 2643, 1987, 987, and 4373, House Committee on Military Affairs, 76th Congress, 1st Session, February-March 1939; U.S. Military Academy, *Strategic and Critical Raw Materials* (West Point: Department of Economics, Government, and History, 1940 and 1944); General Preference Order M-82, January 23, 1942; Jonathan Marshall, *To Have and Have Not: Southeast Asian Raw Materials and the Origins of the Pacific War* (Berkeley: University of California Press, 1995); Alfred E. Eckes Jr., *The U.S. and the Global Struggle for Minerals* (Austin: University of Texas Press, 1979).

15. House Ways and Means Committee, *Taxation of Marihuana,* hearings on HR 6385, 75th Congress, 1st Session, April 27-30 and May 4, 1937; Senate Finance Committee, *Taxation of Marihuana,* hearings on HR 6906, 75th Congress, 1st Session, July 12, 1937.

16. McAllister, "The International Connection"; Eckes, *The U.S. and the Global Struggle.*

17. Robert Marsh, "The Illinois Hemp Project at Polo in World War II," *Journal of the Illinois State Historical Society* 60(4) (1967): 391-410; James F. Hopkins, *A History of the Hemp Industry in Kentucky* (Lexington: University of Kentucky Press, 1951); John Garland, "Hemp: A Minor American Fiber Crop," *Economic Geography* 22 (1946): 126-132.

18. McAllister, *Drug Diplomacy,* Chapter 5.

19. Ibid., Chapter 6.

20. Ibid., Chapter 7; Alfred R. Lindesmith, *The Addict and the Law* (Bloomington: Indiana University Press, 1965); Rufus King, *The Drug Hang-Up: America's Fifty-Year Folly* (New York: Norton, 1972); William Butler Eldridge, *Narcotics and the Law: A Critique of the American Experiment in Narcotic Drug Control* (Chicago: University of Chicago Press, 1962); John F. Galliher, David P. Keys, and Michael Elsner, "Lindesmith v. Anslinger: An Early Government Victory in the Failed War on Drugs," *Journal of Criminal Law and Criminology* 88(2) (1998): 661-683.

21. J. F. Kramer and D. C. Cameron, *A Manual on Drug Dependence* (Geneva: World Health Organization, 1975); P. H. Hughes, K. P. Canavan, G. Jarvis, and A. Arif, "Extent of Drug Abuse: An International Review with implications for Health Planners," *World Health Statistics Quarterly* 36(3-4) (1983): 394-497; World Health Organization, *Youth and Drugs* (Geneva: WHO, 1973); Paul B. Stares, *Global Habit: The Drug Problem in a Borderless World* (Washington, DC: Brookings Institution, 1996); John C. Burnham, *Bad Habits: Drinking, Smoking, Taking Drugs, Gambling, Sexual Misbehavior, and Swearing in American History* (New York: New York University Press, 1993); Giffen, Endicott, and Lambert, *Panic and Indifference;* Virginia Berridge and Betsy Thom, "The Relationship Between Research and Policy: Case Studies from the Postwar History of Drugs and Alcohol," *Contemporary Drug Problems* 21 (1994): 599-629; Philip Bean, *The Social Control of Drugs* (New York: Halstead Press, 1974); Marcel de Kort, "The Dutch Cannabis Debate, 1968-1976," *Journal of Drug Issues* 24(3) (1994): 417-427; Govert Frank van de Wijngaart, *Competing Perspectives on Drug Use: The Dutch Experience* (Amsterdam: Swets and Zeitlinger, 1991); Charles D. Kaplan, "The Uneasy Consensus: Prohibitionist and Experimentalist Expectancies Behind the International Narcotics Control System," *Tijdschrift voor criminology* 26 (1984): 98-109; Hans-Joerg Albrecht and Anton van Kalmthout, *Drug Policies in Western Europe* (Freiburg: Max-Planck-Institut für ausländisches und internationales Strafrecht, 1989); Per Stangeland (Ed.), *Drugs and Drug Control* (Oxford: Oxford University Press/Norwegian University Press, 1987); Ian McAllister, Rhonda Moore, and Toni Makkai, *Drugs in Australian Society* (Melbourne: Longman Cheshire, 1991); M. Z. Khan, *Drug Use Amongst College Youth* (Bombay: Somaiya Publications, 1985); C. Even-Zohar, "Drugs in Israel: A Study of Political Implications for Society and Foreign Policy," in Luiz R. S. Simmons and Abdul A. Said (Eds.), *Drugs, Politics, and Diplomacy: The International Connection* (Beverly Hills, CA: Sage, 1974); Friman, *Narcodiplomacy;* Alfred W. McCoy, *The Politics of Heroin: CIA Complicity in the Global Drug Trade* (New York: Lawrence Hill, 1991); John M. Kramer, "Drug Abuse in the Soviet Union," *Problems of Communism* 37(2) (1988): 28-40.

22. The 1912 Hague Convention, a 1925 Agreement on Prepared Opium that applied to East Asian colonies, the 1925 International Opium Convention, the 1931 Limitation Convention, a 1931 agreement on the suppression of opium smoking

(Bangkok) that applied to East Asian colonies, the 1936 Illicit Trafficking treaty, a 1946 protocol transferring league authority to the UN, the 1948 Synthetic Narcotics Protocol, and the 1953 Opium Protocol.

23. Single Convention on Narcotic Drugs, 1961, United Nations, *Treaty Series,* Volume 520, p. 151 and Volume 557, p. 280.

24. McAllister, *Drug Diplomacy,* Chapter 8.

25. Convention on Psychotropic Substances, 1971, United Nations, *Treaty Series,* Volume 1019, p. 176.

26. William B. McAllister, "Conflicts of Interest in the International Drug Control System," in William O. Walker III (Ed.), *Drug Control Policy: Essays in Historical and Comparative Perspective* (University Park: Pennsylvania State University Press, 1992).

27. See note 21.

28. McAllister, *Drug Diplomacy,* Chapter 9.

29. *Declaration of the International Conference on Drug Abuse and Illicit Trafficking and Comprehensive Multidisciplinary Outline of Future Activities in Drug Abuse Control,* United Nations document ST/NAR/14 (1988); *Systemwide Action Plan,* United Nations documents E/1990/39 (including addenda) and A/RES/S-17/2; Jack Donnelly, "The United Nations and the Global Drug Control Regime," in Peter H. Smith (Ed.), *Drug Policy in the Americas* (Boulder, CO: Westview Press, 1992), pp. 282-304.

Chapter 7

Federal Policy in the Post-Anslinger Era: A Guide to Sources, 1962-2001

Joseph F. Spillane

Harry J. Anslinger retired as head of the FBN in 1962, at age seventy. The singular achievement of his thirty-two-year tenure, as Anslinger himself might have acknowledged, was the maintenance of a remarkably consistent federal drug control philosophy. Anslinger's philosophy was not original, for it had already been articulated at the federal level years before he took office. In fact, there were four pillars of federal policy, each of which had cultural roots that predated any formal federal drug control. First, federal policy assumed that the use of opiates, cocaine, and (later) marijuana produced degenerative effects in their users. These effects were always present in addicts, thus rendering drug takers socially unproductive. Worse, users were a danger and a menace to legitimate society. These drug effects made the addicts' problem a public one that demanded a response in social policy. The second assumption of federal drug policy was that the cornerstone of effective control was the deterrent power of the criminal law and its strict enforcement. Third, federal policy took for granted that illicit drug use was largely a problem of willful personal behavior and that drug habits were simply habits, not the manifestation of a diseased state requiring therapeutic intervention. To be sure, the opening of federal narcotics hospitals in the 1930s suggested some version of a medical model. Closer examination reveals that the hospitals were largely predicated on the idea of keeping the addict away from the supply, a goal not inconsistent with the Anslinger model. Finally, the fourth pillar of the federal drug program was based on the assumption that effective policy required the maintenance of the orthodox position. Dissent was intolerable not merely because it challenged the view of authority, but because the toleration of dissent would actually produce a more severe drug problem. Like his counterparts before and after, Anslinger believed in the power of a moral message to shape public behavior; to undermine the message was to push public behavior in the wrong direction.

Four decades have passed since Anslinger's retirement. One could argue that little has happened since 1962 to alter the fundamental character of drug control in the United States. Certainly it is not hard for observers of contemporary drug policy to find evidence that the "four pillars" still command powerful support. Moreover, the cultural backdrop that gave Anslinger's message such resonance remains in place. Much of what existed in 1919 exists today. Indeed, one of the underlying ideas of this book has been to expose the historical roots of a drug control regime that continues to thrive in a new century. America's longest war goes on—those who forget the past may be doomed to perpetuate it.

Despite this harsh reality, there has been a great deal of change in the post-Anslinger decades. The war on drugs has grown larger and vastly more complex than it was in Anslinger's day. Medical models of addiction have moved back into federal policy, where they maintain an uneasy coexistence with enforcement-oriented programs. Although it is beyond the scope of this book to fully document these changes, what follows is a bibliographic guide for readers interested in pursuing the story of federal drug control from Anslinger to 2001. This chapter sketches some of the major developments of recent drug control history and identifies selected sources that offer further information and insight.

SEEDS OF CHANGE

As others have observed, the maintenance of policy "orthodoxy" by Anslinger and the FBN was never automatic nor easy. The events of 1919 did not silence forever the critics of federal drug policy. After World War II, the intellectual and medical critics of American drug programs gained considerable prominence. Although they wandered in the policy wilderness during Anslinger's tenure, by 1962 it seemed increasingly obvious that the critics' message would begin to influence federal drug strategies.

The social science fields were fertile ground for researchers interested in the problems of urban America, race, crime, and juvenile delinquency. The heroin epidemic that many cities experienced after World War II suggested to some researchers that drug use and, by implication, drug policy lay at the intersection of these complex social concerns. The influential literature in this area includes: Isidor Chein, D. L. Gerard, R. S. Lee, and Eva Rosenfeld, *The Road to H: Narcotics, Delinquency, and Social Policy* (New York: Basic Books, 1964); Kenneth B. Clark, *Dark Ghetto: Dilemmas of Social Power* (New York: Harper and Row, 1965); Harold Finestone, "Cats, Kicks, and Color," *Social Problems* 5 (1): 3-13, 1957; and HARYOU [Harlem Youth Opportunities Unlimited], *Youth in the Ghetto: A Study of the Consequences of Powerlessness and a Blueprint for Change* (New York: Harlem Youth Opportunities Unlimited, 1964). A good deal of the intellectual foundation and some of the research for these works predated Anslinger's retirement but, as the publication dates suggest, public discussion became widespread in the 1960s.

One of the best markers of the growing influence of new ideas on federal drug policy is the *President's Commission, Proceedings of the White House Conference on Narcotic and Drug Abuse, Final Report* (Washington, DC: GPO, 1962). The decision to convene a conference on drug abuse was seen by Anslinger, correctly, as a repudiation of his policies by the Kennedy administration. Another interesting marker of change at the executive level is the *Task Force Report: Narcotics and Drug Abuse, President's Commission on Law Enforcement and Administration of Justice* (Washington, DC: GPO, 1967).

The social science literature just cited tended to offer indirect critiques of federal drug policy. The end of the Anslinger regime also provided an opportunity for many of his most direct critics. Publication of the American Bar Association-American Medical Association Joint Committee on Narcotic Drugs, *Drug Addiction: Crime or Disease?* (Bloomington: Indiana University Press, 1961) provided a forum for many prominent Anslinger-era critics, including Alfred Lindesmith, Rufus King, and Morris Ploscowe. For some representative examples of this critical literature, see: Alfred R. Lindesmith, *The Addict and the Law* (New York: Vintage Books, 1965); Rufus King, *The Drug Hang-Up: America's Fifty-Year Folly* (New York: W.W. Norton, 1972), and Edwin M. Schur, *Crimes Without Victims* (Englewood Cliffs, NJ: Prentice-Hall, 1965).

Two works capture nicely for the modern reader the experience of dissent. One might begin with Nat Hentoff, *A Doctor Among the Addicts* (New York: Rand McNally, 1968), an account of the work of Dr. Marie Nyswander. She was not only a dissenter, based in part on her experience working at the federal narcotics hospital at Lexington, Kentucky, but was, together with Vincent P. Dole, the pioneer of methadone maintenance for heroin addicts. For a more academic and very interesting treatment, see David Patrick Keys and John F. Galliher, *Confronting the Drug Control Establishment: Alfred Lindesmith As a Public Intellectual* (Albany, NY: State University of New York Press, 2000). This last book should be required reading for those wishing to understand the efforts of Anslinger and the FBN in trying to quash dissent.

THE PSYCHOACTIVE REVOLUTION

Federal drug policy in the 1960s confronted a new landscape of drug use. The focus of drug control efforts had, for decades, been on the opiates, cocaine, and marijuana. By the end of World War II, however, it was already obvious that new drugs of abuse were emerging. The popularization of the amphetamines and the barbiturate drugs brought new forms of drug abuse that existing federal law did not anticipate. Indeed, Anslinger and the FBN engaged in a vigorous effort to keep these substances outside the purview of the federal drug enforcement apparatus. So long as these stimulant and depressant drugs enjoyed widespread medical use, the federal government was wary of control efforts. Any attempts at control that were made, as John Swann has observed in this volume, was the work of the FDA, not the FBN.

By the 1960s, misuse of amphetamines and barbiturates reached greater levels than ever before. More important, the problems of "pill popping" became firmly linked, in the public mind, with an emergent youth drug culture. At the same time, the so-called hallucinogenic drugs generated a storm of media coverage, especially LSD. The creation of the Bureau of Drug Abuse Control (BDAC) within the FDA in 1965 significantly expanded federal authority over all drugs with potential for abuse. BDAC was later merged with the FBN into the Bureau of Narcotics and Dangerous Drugs (BNDD) in 1968, bringing the control of traditional "narcotic" drugs under the same federal agency as that for psychoactive substances.

Since then new drugs of abuse have appeared at a steady rate, requiring a continuous process of evaluation and reevaluation of the abuse potentials of widely used drug products. This has, in turn, generated a large and very useful body of scholarly literature. One of the first major works of drug history that covers these early battles is Edward Brecher, *Licit and Illicit Drugs* (Boston: Little, Brown, 1972). Specific studies of amphetamine issues include David E. Smith, Donald R. Wesson, Millicent E. Buxton, Richard B. Seymour, J. Thomas Ungerleider, John P. Morgan, Arnold J. Mandell, and Gail Jara (Eds.), *Amphetamine Use, Misuse, and Abuse* (Boston: G.K. Hall, 1979) and Lester Grinspoon and Peter Hedblom, *The Speed Culture* (Cambridge, MA: Harvard University Press, 1975). There have been many congressional hearings on this subject, but the one that best demonstrates the extension of federal drug control efforts is the House Select Committee on Crime, *Crime in America: Why Eight Billion Amphetamines?* Hearings, 91st Congress, 1st session (Washington, DC: GPO, 1970).

A variety of substances have their own interesting studies. LSD has been the subject of many useful works. One might begin by reading the interesting social history by Martin A. Lee and Bruce Shlain, *Acid Dreams* (New York: Grove Press, 1992). A major text on quaaludes is Mathea Falco, *Methaqualone: A Study of Drug Control* (Washington, DC: Drug Abuse Council, 1975). For coverage of the benzodiazepines, of which Valium is perhaps the best known, readers should see Mickey C. Smith, *Small Comfort: A History of the Minor Tranquilizers* (New York: Praeger, 1985). In the 1980s, the emergence of MDMA (Ecstasy) prompted still another battle over the appropriate level of federal control. This episode is recounted in Jerome Beck and Marsha Rosenbaum, *In Pursuit of Ecstasy* (Albany, NY: State University of New York Press, 1994).

For an interesting perspective on the process of defining new drug problems, readers should see Phillip Jenkins, *Synthetic Panics: The Symbolic Politics of Designer Drugs* (New York: New York University Press, 1999).

THE NIXON ADMINISTRATION

If the seeds of change had been planted prior to the 1968 election of Richard Nixon as president of the United States, it must be said that the Nixon administration (1969-1974) brought to fruition a recognizably modern drug control regime. Pushed by the political concerns raised by a rising tide of recreational drug use among younger Americans, a serious heroin epidemic in many major cities, and worries about heroin use among Vietnam servicemen, the Nixon White House took federal drug policy in new directions.

On the legislative front, the Comprehensive Drug Abuse and Control Act of 1970 combined and replaced all the previous federal legislation relating to narcotic drugs, including the 1914 Harrison Narcotics Act. Title II of this legislation, the Controlled Substances Act, created five drug control categories, known as schedules, into which controlled substances would be placed, based on their harmfulness, potential for abuse, and medical utility. At the executive level, the Nixon administration created the Special Action Office for Drug Abuse Prevention (SAODAP). The mission of SAODAP was to formulate an overall national drug control strategy, to organize and coordinate the responses of multiple federal agencies involved with the drug issue, and to oversee an ambitious program of federally funded drug treatment. Another executive agency, the Office of Drug Abuse Law Enforcement (ODALE) was charged with increasing the street-level "war" on drug distribution. The administrative reorganization of federal drug control efforts culminated in 1973 with the creation of the Drug Enforcement Administration (DEA) from the merger of the BNDD and the ODALE. That same year, numerous treatment and prevention programs were consolidated with the creation of the National Institute on Drug Abuse (NIDA).

One major study on the Nixon administration's drug policies is the short, scholarly review by Elaine B. Sharp, *The Dilemma of Drug Policy in the United States* (New York: Harper Collins, 1994). Two general histories of U.S. drug control efforts devote extensive time to the Nixon era. Jill Jonnes' book, *Hep-Cats, Narcs, and Pipe Dreams: A History of America's Romance with Illegal Drugs* (New York: Scribner's, 1996), pays close attention to the context of the Nixon administration's efforts, as does the updated version of David F. Musto's classic text, *The American Disease: Origins of Narcotic Control,* Third Edition (New Haven: Yale University Press, 1999).

The first generation of writers who offered retrospective accounts of the Nixon-era antidrug strategies, no matter how they differed in their policy positions, did not regard highly the policy developments of this era. Among the most notable of these works are: David J. Bellis, *Heroin and Politicians: The Failure of Public Policy to Control Addiction in America* (Westport, CT: Greenwood Press, 1981); Edward Brecher, *Licit and Illicit Drugs* (Boston: Little, Brown, 1972); Edward Jay Epstein, *Agency of Fear: Opiates and Political Power in America,* Revised Edition (New York: Verso, 1990); John Kaplan, *The Hardest Drug: Heroin and Public Policy* (Chicago: University of Chicago Press, 1983); and Arnold S. Trebach, *The Heroin Solution* (New Haven: Yale University Press, 1982).

Recent years have seen a significant amount of revisionist work on this era. In 1998, Yale University sponsored the Conference on One Hundred Years of Heroin. Organized by David Musto, the meetings brought together many of the major figures of Nixon-era drug policy for a largely positive assessment of the work that had been done. Papers from that meeting appear in published form in David F. Musto and Pamela Korsmeyer (Eds.), *One Hundred Years of Heroin* (Westport, CT: Auburn House, 2002). Another recent work that focuses on political developments in the Nixon years is David F. Musto and Pamela Korsmeyer, *The Quest for Drug Control: Politics and Federal Policy in a Period of Increasing Substance Use* (New Haven: Yale University Press, 2002). Musto and Korsmeyer's study focuses on the period 1960-1980, and draws upon presidential records to describe the internal policy debates in successive administrations. Readers might also note that this work includes a searchable CD-ROM of the documents the authors employed in their study. Michael Massing's *The Fix* (New York: Simon and Schuster, 1998), describes the Reagan-Bush-Clinton drug war as an abject failure, but looks back to Nixon administration strategies for a model of what might be done with contemporary drug policy.

INTERNATIONAL DRUG ENFORCEMENT

One of the most striking changes since Anslinger's retirement has been the increasing scope and scale of international drug enforcement. The FBN was the first federal agency to field foreign agents in the service of drug enforcement, but their presence was very limited. A small number of FBN agents roamed the drug trafficking regions largely on their own, attempting to disrupt distribution networks wherever possible. Many of these agents were extremely resourceful independent operators, and their adventures often remarkable, but they did not constitute a systematic international control effort.

Since the FBN ceased to exist, and especially after the creation of the DEA in 1973, international drug policing has grown in unprecedented ways. The place to begin examining this work is with Ethan A. Nadelmann, *Cops Across Borders: The Internationalization of US Criminal Law Enforcement* (University Park: Penn State University Press, 1993). Nadelmann's work places the DEA in a broader historical context, and offers the reader a wealth of insights into the nature of international drug enforcement. There is, as yet, no general history of the DEA's international drug control activities. For specific aspects of international operations, see the following sampling of works: Alfred R. McCoy, *Politics of Heroin: CIA Complicity in the Global Drug Trade* (New York: Lawrence Hill Books, 1991); Peter Reuter, *Quest for Integrity: The Mexican-U.S. Drug Issue in the 1980s* (Santa Monica, CA: RAND Corporation, 1992); Peter Dale Scott and Jonathan Marshall, *Cocaine Politics* (Berkeley: University of California Press, 1991); and William O. Walker, *Drug Control in the Americas* (Albuquerque: University of New Mexico Press, 1981) and *Opium and Foreign Policy* (Chapel Hill: University of North Carolina Press, 1991).

The internationalization of drug enforcement has seemed characteristic of globalization processes more generally. One study that considers this question is Paul B. Stares, *Global Habit: The Drug Problem in a Borderless World* (Washington, DC: Brookings Institution, 1996). Readers interested in the issue of drug enforcement and the extent and limits of national sovereignty should see Maria Celia Toro, "The Internationalization of Police: The DEA in Mexico," *Journal of American History* 86 (September) (1999): 623-640.

MARIJUANA

The reconsideration of federal drug legislation and policy during the Nixon administration also included some support for marijuana law reform at the federal level. The 1937 Marihuana Tax Act had come under criticism in 1969 when the U.S. Supreme Court had reversed the conviction of Timothy Leary, in part because of their view that the 1937 act was constitutionally suspect. More important, the increase in marijuana use among many segments of American society prompted concerns over what many saw as a huge disparity between the physical/psychological harms supposedly caused by marijuana and the actual punitive sanctions attached to the marijuana laws. The Nixon-appointed National Commission on Marihuana and Drug Abuse, First Report, *Marihuana: A Signal of Misunderstanding* (Washington, DC: GPO, 1972) provides an overview of the changes in the "official" story of marijuana. Although the commission sought to discourage marijuana use, its conclusions stated, "criminalization of possession of marihuana for personal use is socially self-defeating as a means of achieving this objective" (p. 147). Similar conclusions were reached in the Domestic Council Drug Abuse Task Force, *White Paper on Drug Abuse* (Washington, DC: GPO, 1975). This report, presented to President Gerald Ford in 1975, echoed some of the reformist language in dealing with marijuana.

Efforts to reform federal drug laws culminated during the Jimmy Carter administration (1977-1981). A discussion of these developments can be found in Patrick Anderson, *High in America: The True Story Behind NORML and the Politics of Marijuana* (New York: Viking Press, 1981). Anderson's work describes in some detail the movement for marijuana decriminalization at the outset of Carter's presidency in 1977 and the abrupt failure of those efforts just over one year later. One source for the failure of marijuana law reform was a vigorous "parents' movement" that opposed any liberalization of drug laws and urged those who saw marijuana as essentially harmless to reconsider their views. The parents' movement, discussed in the work of Jonnes, Massing, and Musto, was essentially the resurgence of the popular temperance sentiment upon which the drug war had been based for so long. This modern manifestation has not been thoroughly examined by historical scholarship. Among the best representative examples of the parents' movement argument against marijuana and marijuana law reform are Marsha Manatt, *Parents, Peers and Pot* (Washington, DC: NIDA, GPO, 1979) and Peggy Mann, *Marijuana Alert* (New York: McGraw-Hill, 1985).

A more recent and very useful work that deals with many aspects of marijuana's history is Lester Grinspoon and James B. Bakalar, *Marijuana: The Forbidden Medicine* (New Haven: Yale University Press, 1993).

COCAINE AND DRUG WAR ESCALATION

There is universal agreement that at some point after the election of Ronald Reagan in 1980 the United States embarked on a "new" drug war—new in scope, at least, if not in substance. Exactly when this new drug war was launched remains open to debate. Some accounts begin with the collapse of Carter's drug law reform proposals, the parents' movement, and Reagan's election. Others refer to 1982, when the Department of Defense Authorization Act of 1982 amended the century-old Posse Comitatus Act to allow U.S. military resources to be devoted to drug control and interdiction. Still others point to the popular antidrug campaigns to which Nancy Reagan devoted herself, starting in 1982 and featuring the national "Just Say No" effort in 1985. Finally, others claim that the emergence of crack cocaine altered the national antidrug priorities, resulting in the Anti-Drug Abuse Act of 1986. This legislation further expanded the military role in federal drug control, increased funding for the control of drug smuggling, authorized the president to "certify" whether source countries were cooperating in antidrug efforts, created mandatory minimum prison terms for drug offenses, and increased spending for prisons and drug treatment. The last of these provisions was important in stopping a decline in federal spending on treatment, but was only a small part of a legislative package that overwhelmingly returned federal policy to the drug enforcement/supply control position. In terms of actual drug programming, this last piece of legislation may be the best marker of change, although the forces behind it were set in motion earlier. The spirit of this renewed drug war is perhaps best articulated in the Office of National Drug Control Policy, *National Drug Control Strategy* (Washington, DC: GPO, 1989). For an overview of the scale and scope of today's federal effort, see Patrick Murphy, Lynn E. Davis, Timothy Liston, David Thaler, and Kathi Webb, *Improving Anti-Drug Budgeting* (Santa Monica, CA: RAND Corporation, 2000).

Little historical research is available on this period, though it surely deserves attention. Professional historians have remained fixated on the Progressive Era roots of American drug policy; however, a number of academics present at the "front lines" have begun to offer retrospective accounts. Craig Reinarmann and Harry G. Levine (Eds.), *Crack in America: Demon Drugs and Social Justice* (Berkeley: University of California Press, 1997), is an extraordinary, historically conscious look at the experience of the 1980s in federal drug control policy, though strictly from a dissenter's point of view. Another useful source for information is Steven Wisotsky, *Beyond the War on Drugs* (Buffalo: Prometheus Books, 1990). Among the best journalistic accounts of this period is Dan Baum, *Smoke and Mirrors: The War on Drugs and the Politics of Failure* (Boston: Little, Brown, 1996).

There is a fine series of scholarly works that confronts drug policy past, present, and future. These studies include Elliott Currie, *Reckoning: Drugs, the Cities, and the American Future* (New York: Hill and Wang, 1993); Mathea Falco, *The Making of a Drug-Free America* (New York: Times Books, 1994); Mark A. R. Kleiman, *Against Excess: A Drug Policy for Results* (New York: Basic Books, 1992); and Franklin Zimring and Gordon Hawkins, *The Search for Rational Drug Control* (Cambridge, England: Cambridge University Press, 1992).

Finally, two 2001 works should be required reading for any serious student of U.S. drug policy. David T. Courtwright, *Forces of Habit: Drugs and the Making of the Modern World* (Cambridge: Harvard University Press, 2001) is a remarkable survey of the global history of drugs, both licit and illicit. Courtwright explores the ways in which certain substances become popular commodities, why some of these eventually become the objects of control efforts, and why these control efforts succeed or fail. Robert J. MacCoun and Peter Reuter, *Drug War Heresies: Learning from Other Vices, Times, and Places* (Cambridge, England: Cambridge University Press, 2001) is, like Courtwright's work, broadly comparative and multidisciplinary. The authors draw upon a variety of regulatory experiences to offer suggestions about the consequences of various drug policy choices.

URLs FOR PRIMARY DOCUMENTS

Reorganization Plan No. 1 of 1968
<http://www.access.gpo.gov/uscode/title5a/5a_4_86_2_.html>

Legal References on Drug Policy, Federal Court Decisions on Drugs by Decade, 1960
<http://www.druglibrary.org/schaffer/legal/legal1960.htm>

Legal References on Drug Policy, Federal Court Decisions on Drugs by Decade, 1970
<http://www.druglibrary.org/schaffer/legal/legal1970.htm>

Legal References on Drug Policy, Federal Court Decisions on Drugs by Decade, 1980
<http://www.druglibrary.org/schaffer/legal/legal1970.htm>

Marihuana: A Signal of Misunderstanding, the Report of the National Commission on Marihuana and Drug Abuse, March, 1972
<http://www.drugtext.org/library/reports/nc/ncmenu.htm>

Special Action Office for Drug Abuse Prevention, 1972
<http://www4.low.cornell.edu/uscode/21/1111-1114.html>

The Drug Hang-Up: America's Fifty-Year Folly by Rufus King, 1972
<http://www.druglibrary.org/special/king/dhu/dhumenu.htm>

"Drive to Curb Hard Drugs Gets a No. 1 Priority," *U.S. News and World Report,* April 3, 1972
<http://www.druglibrary.org/schaffer/history/e1970/usnews1.htm>

The Facts About Drug Abuse, The Drug Abuse Council, 1980
<http://www.drugtext.org/library/reports/fada/fadamenu.htm>

Anti-Drug Abuse Act of 1986
<http://www.usembassy-mexico.gov/bbf/bfantiab.htm>

Psychoactive Substances and Violence by Jeffrey A. Roth, Research in Brief, U.S. Dept. of Justice, February 1994
<http://www.druglibrary.org/schaffer/GovPubs/psycviol.htm>

Government Publications on Drugs and Drug Policy
<http://www.druglibrary.org/schaffer/GovPubs/GOVPUBS.HTM>

Office of National Drug Control Policy, The White House
<http://www.health.gov/NHIC/NHICScripts/Entry.cfm?
HRCode+HR2641>

Index

Page numbers followed by the letter "i" indicate illustrations; those followed by the letter "n" indicate notes; those followed by the letter "t" indicate tables.